RADIO
FREE
BOSTON

RADIO FREE BOSTON

the rise and fall of

NORTHEASTERN UNIVERSITY PRESS BOSTON

Northeastern University Press
An imprint of
University Press of New England
www.upne.com
© 2013 Carter Alan
Manufactured in the
United States of America
Designed by Mindy Basinger Hill
Typeset in Calluna Pro

For permission to reproduce
any of the material in this book, contact Permissions,
University Press of New England, One Court Street, Suite 250,
Lebanon, NH 03766; or visit www.upne.com.

Library of Congress Cataloging-in-Publication Data
Alan, Carter.
Radio free Boston : the rise and fall of WBCN / Carter Alan ;
foreword by Steven Tyler.
pages   cm.
ISBN 978-1-55553-729-6 (pbk. : alk. paper) —
ISBN 978-1-55553-826-2 (ebook)
1. WBCN (Radio station : Boston, Mass.)—History.
2. Music radio stations—Massachusetts—Boston—History. I.
Title.
PN1991.67. M86A43 2013
384.54'53—dc23   2013004735

FOR CARRIE

contents

Color plates follow page 158.

foreword

STEVEN TYLER

Imagine a time when you could get good in a band by practicing in a basement or garage, when you could play in local bars or clubs, and when local
radio stations were actually running themselves. Real human personalities
owned the music as much as the bands that wrote it; they frequented
clubs, knew personally the pulse of local talent, and knew the goings-on
of a community or city. This, in fact, is how everyone from Willy Dixon to
Elvis, the Beatles, the Beastie Boys, Run DMC, and a little unknown band
from Boston called Aerosmith got their foot in the door and proceeded to
wreak havoc and bring the house down for the next forty years.

It's all Maxanne Sartori and WBCN's fault. We were ruffling our feathers
one night when we caught her attention at a club called Paul's Mall. And
before you knew it, I was sitting on her lap in the studio at 5:00 a.m. on
a Sunday morning doing a cat-and-dog public service announcement. I
tickled her fancy, so to speak, into playing "Dream On" and turning the
daytime jocks on to our demos, which we had just recorded at Intermedia
Studios on Newbury Street.

The timing was perfect, and the rest is history.

Due to the passion of these local DJs, we were able to ignite a fire that
would burn in the hearts of millions . . .

THANK YOU WBCN,
Steven

preface

Just weeks after WBCN folded in August of 2009, I got a call from Stephen Hull from University Press of New England. "Would I be interested in writing a book telling legends and fables from the birthplace of Boston progressive radio? My qualifications were that I'd actually been employed as a DJ and music director at WBCN for a good chunk of its forty-one-year history (from 1979 to 1998) and that I'd already published a couple of books. Although the idea intrigued me, the thought of militantly organizing my writing time around an already-bulging schedule at 'BCN's sister station WZLX did not intrigue me. I had jousted with that stress monster before, and as Confucius or maybe Charles Laquidara said, all work and no play leaves one a bundle of nerves and the master of nothing. So I stonewalled, fascinated by Hull's idea while equally fearing it, before declining. But Stephen refused to close the door and urged me to think about it.

Enter the wife here. Carrie also worked at WBCN, as a sales secretary from 1989 until 1995 (conveniently leaving the station moments after we tied the knot). She argued, "You know most of the people that worked there; why shouldn't you write it?" That was logical, but the issue remained that accepting the project would result in no life for an extended period of time, like when you were in fifth grade and all your friends ran past the house in a raging snowball fight, but your mom wouldn't let you suit up and join them until you finished that damn book report on *Robinson Crusoe*. Multiply that example into a year or two of abstention. Ouch!

"Well, if you don't write it, someone else who didn't work there will do it . . . like that guy who wrote the Aerosmith book."

"Stephen Davis," I replied. "He also wrote *Hammer of the Gods—the* Led Zeppelin book, as well as one on Jim Morrison, Bob Marley, Guns and Roses . . ."

"Yeah, Stephen Davis will write it. Then what?" Now, those are fighting words because I'm a bit envious of Davis. He's a damn good writer and, along with Cameron Crowe, seems to have had the greatest rock and roll journalism adventures on record. I mean, he knows the truth behind the "Mudshark" mythology! Button pushed, I replied, "Yeah, why should he get the 'BCN story?" I found myself shouting: "He gets all the other stories! I'll do it!"

A callback to Stephen Hull and my fate was set, for not one, or even two, but over three years. Writing in a limited time frame from five to eight every weekday morning with as much time as possible on the weekends, and recording and transcribing over one hundred interviews (104 to be exact, and no, I didn't try to do that), sent the project into serious overtime. I actually tried to bail at one point because my health started to go downhill, but Hull gave me slack and urged me to stay on it, albeit at a less frenetic pace. With apologies to Stephen Davis, I found myself becoming jealous of the fact that he could exclusively write for his supper and not need to divide his time between writing and work. It became a no-win every morning: if I was writing well, I'd be upset that I had to put the computer away and do battle on the Mass Pike. If the writing resembled caveman gibberish, I'd have no further time to wrestle the words into a decent representation of the English language, and I'd hit the shower in frustration.

Nevertheless, commissioned to my task, I found myself lucky and privileged to be able to share the misery of regimentation that all writers must endure. I also feel honored to have been one of the illustrious staff at WBCN and part of a great radio experiment that began in 1968 and survived through some of the most turbulent changes in all of human history. At its best, WBCN represented what a community of believers, not preoccupied with testing the limits of personal gain, could accomplish. At its worst, the station found itself swept out of that humble place and down the inevitable road of capitalism, losing its innocence as it slid toward a desired stock price at the end of the rainbow. WBCN went from Baltic Avenue to Boardwalk with hotels on it, and it wasn't a pretty end, but there was magic created all along that forty-one-year ride past "Go." That's why this book is here: to pull out some of that magic before we all begin to forget it.

WBCN became the major force in Boston radio and, along with WRKO and WXKS (KISS 108), dominated every other station in awareness in the market. Sure, others made that dent occasionally, WCOZ, WEEI, WAAF, and WFNX, but they never approached the status and power of the WBCN legend or became internationally recognized as a true phenomenon of American radio. Recognized in the Rock and Roll Hall of Fame, this is the station that introduced new ideas that became fresh trends, and then accepted dogma. The 'BCN jocks were test pilots, really, getting out there like Chuck Yeager in a leaky X-1, beating the speed of sound and passing the how-to-do-it on to the next generation. Underground, progressive, AOR (album-oriented rock) radio, and alternative were format labels taped on 'BCN's journey through the years, a swath of influence and innovation that dated back to Lyndon Johnson's presidency. Refined in the fires of the sixties antiwar movement, swept into the color-splashed, pixilated eighties world of MTV, and eventually drained of its blood in the consolidated radio industry of the new century, this is the station that rocketed through it all. There are some stations with colorful stories, but none as vibrant as the one you're about to read. WBCN began as a spark on a windy day, somehow catching into a decades-long conflagration. Eventually, that fire would eat itself, but happily, it took a long time . . .

thanks

Johnny and Beth A., Bill Abbate, Katy Abel, Adam 12, Carrie Alan, Dan Beach, Andy Beaubien, Tony Berardini, David Bieber, Bill Bracken, John Brodey, Larry Bruce, Julie Brummer, Bo Burlingham, Cali Calandrello, Mark Cappello, CBS Radio, Nik Carter, Tony Chalmers, Lauren Chiaramonte, Lenny Collins, Mike Colucci, Tom Couch, Steve Crowley, Charles Daniels, Ron Della Chiesa, Deke Diedricksen, Michael Dobo, Matt Dolloff, Kaye Dudek, Billy Finnegan, Michael Fremer, John Garabedian, Clint Gilbert, Jerry Goodwin, Roger Gordy, Andrew Govatsos, Leo Gozbekian, Tommy Hadges, Donna Halper, Mark Hamilton, Mark Hannon, Hardy, Larry Harris, Tami Heide, Stephen Hull, Susan Hunter, Sal Ingeme, Bradley J., Juanita, Bill Kates, Mark Kates, Frank and Janice Kearns, Charlie Kendall, Martin Kessel, Bruce Kettelle, Andrew King, Sam Kopper, Jonathan Kraft, Bob Kranes, Heidi La Shay, Charles Laquidara, Kathryn Lauren, John and Sandra Laurenti, Don Law, Dave Lawrence, Lee Leipsner, Jonathan Lev, Maurice Lewis, Bill Lichtenstein, Larry "Chachi" Loprete, Karalyn Mallozzi, Dan Mason, Dan McCloskey, Paul McGuinness, Bob Mendelsohn, Mim Michelove, Marc Miller, Mark Mingels, Bruce Mittman, Tim Montgomery, Roger Moore, John Mullaney, Patrick Murray, Neil Napolitano, Steve Nelson, Chuck Nowlin, Albert O., Dan O'Brien, Oedipus, Tom O'Keefe, Mark Parenteau, Jim Parry, Al Perry, Rachel Phillips, Dave Pierce, Ron Pownall, Gina Preziosi, Curtis Raymond, Ray Riepen, Tracy Roach, Neal Robert, Tammy Robie, Joe Rogers, Kate Curran Rooney, Stu Rosner, Cathy Rozynek, Jefferson Ryder, Tom Sandman, Don Sanford, John Scagliotti, Matt Schaffer, Danny Schechter, John Sebastian, Steve Segal, Paul "Tank" Sferruzza, Bob Shannon, Ken Shelton, Eli Sherer, Rich Shertenlieb, Shred, Matt Siegel, George Skaubitis, Clark Smidt, Jeannie Smith, Joe Soucise, Sue Sprecher, Bill Spurlin, Tod Stevens, Greg Strassell, Steve Strick, "Mr. Mike" Symonds, Fred Taylor, Melissa Teper, Mike Thomas, Fred Toucher, Lisa Traxler, Steven Tyler, Debbie Ullman, Dinah Vaprin, Mike Ward, Billy West, Sherman Whitman, John Taylor Williams, Gahan Wilson, Norm Winer, Paul Winters, Dave Wohlman, Peter Wolf, Anngelle Wood, WZLX-FM, Kenny Young, and Dana Zazinski. Special thank you to Lord God for bringing it home!

RADIO
FREE
BOSTON

THE AMERICAN REVOLUTION

Joe Rogers had a new job. It didn't pay much, but that wasn't the point. This was a gig to dream about, like scoring in Vegas for a cool million or winning a gleaming new '68 Camaro SS in a raffle. Barely three weeks earlier, the Tufts student had been spinning his thoughtful arrangement of folk and blues records at WTBS-FM in Cambridge, the noncommercial station at MIT, when he'd met a most unusual figure. Filled with energy and ideas, the dynamo of a man radiated a wild entrepreneurial spirit and confidence that made it seem like his crazy schemes could actually work. His name was Ray Riepen, and he came off like some older midwestern lawyer: starched and unflappable, with the bulletproof assurance of an innocent newly arrived in the scheming Versailles of sophisticated Bostonians, and itching to work his way into the next hand. Even at Rogers's relatively young age, he had a sense that the world was often unkind to people with dreams; yet, this guy glowed with foresight and just seemed to be a heck of a lot smarter than anybody he'd ever met.

So, Rogers bought Riepen's scheme; what did he have to lose anyway? It wasn't like he had some big-time radio career to worry about; the college kid at Tufts merely wanted to stay out of the army and listen to music. Certainly not a professional disc jockey like Alan Freed or Arnie Ginsburg, he had no desire to be some fast-talking AM radio jive ass. A natural talent on the air, Rogers loved to blend the many artists and styles plucked from his treasured record collection for the enjoyment of his audience. Now, barely a month later, he had gotten the phone call from Riepen about that crazy idea they'd talked about. This was actually going to happen! Gathering up the records he needed was a deliciously unhurried process as the DJ lingered in the moment, anticipating the moods he would create, and the cuts, intros, and segues he would perform. Then, hefting up the heavy crate of vinyl, Rogers headed toward Back Bay and the 171 Newbury Street studios of WBCN-FM, the flagship station of the Boston Concert Network and bastion of classical music programming in the Northeast. Tonight, 15 March 1968, there'd be some new sounds coming out of the tower, judging by the records Rogers carried: *We're Only in It for the Money* by Frank Zappa and the Mothers of Invention; some Elmore James; *Boogie with Canned Heat*; *Papa's Got a Brand New Bag* from James Brown; *Buffalo Springfield Again*; a Kweskin Jug Band disc; and *Fresh Cream* from recently departed Bluesbreaker Eric Clapton and his two new bandmates.

Rogers reached 171 Newbury, went in the front door, and took the battered, old elevator up to the third floor, relieved to put his heavy crate of records down. The doors opened into WBCN's cramped office space, now darkened, as it should be this late after a long workweek. The place seemed deserted. The main air studio was up a couple levels beyond the reach of the elevator, so he lifted his burden again and hiked up the timeworn stairs to the attic. It was cramped up there: a main room barely twelve feet across with another production studio and a massive classical music library all squeezed into a tiny loft. As he crested the steps he saw some figures milling about: Riepen; a guy he recognized from the Boston Tea Party named Don Law; fellow college radio jocks Tommy Hadges and Jack Bernstein; and his engineer buddy Steve Magnell from Tufts. Rogers also saw the man he'd be relieving sitting at the control board. He smiled at him nervously, but the man grinned back. Although Ron Della Chiesa had become a veteran announcer and the program director of WBCN, he remembered the excitement of his first show at the station eight years earlier on Christmas Day, and he knew how Rogers felt.

Do you remember the front cover of the Pink Floyd album *Wish You Were Here?* Two men in business suits stood in a studio lot shaking hands—and one of them had burst into flames. This was that kind of moment, the two DJs standing on opposite sides of a musical chasm. Chiesa lived and breathed classical music and as a result had forged a steady upward path at the Boston Concert Network soon after graduating from Boston University. Rogers's musical passions traveled elsewhere, in American blues and traditional folk; his enthusiasm embraced such musical outsiders as the Mothers of Invention, the Fugs, and his beloved Holy Modal Rounders. Chiesa's smooth and debonair on-air delivery was a far cry from Rogers's somewhat timid approach, his college-radio experience never demanding he develop any amount of polish or panache. But, the younger, scruffier DJ didn't care much about his voice or style, just like he didn't care so much about how he dressed. Rogers preferred to speak on the radio *through* his music. And so he would tonight.

Although it didn't seem like such a big deal at the time, Joe Rogers's first shift at WBCN became a shot heard round the world or (at least) a crate of tea tossed into Boston Harbor. Time and hindsight would be the forces that eventually heaped great significance on this moment. It was, as Ray Riepen surmised, the time of a new American revolution. The rebellions in cultural, sexual, musical, artistic, intellectual, and political freedoms that began slowly at the end of the fifties had gathered speed in the early sixties and burst into full flower by the middle of the decade. The youth of America and free Europe, witnessing the daily horrors and rising body counts broadcast every night in living color from Southeast Asia, united in a bond as strong as those that had toppled totalitarian regimes in the past. All the turbulence of a new generation unsullied by convention, heedless of authority, and joyfully seeking answers had arrived on this night in a most unusual place: the control room of a classical music FM radio station. Even though Ron Della Chiesa hadn't yet turned thirty and was quite sympathetic to the forces of change, he still represented an old guard about to be swept away.

Rogers settled into the studio chair, assuming his bluesy radio identity as "Mississippi Harold Wilson," and then cued up a record while waiting for Chiesa's final selection to wind down. It was almost 10:00 p.m. Time had begun to accelerate in the last few moments, like it usually did before stressful and important occasions, as Mississippi felt the excitement of leaving his ten-watt college radio world behind to sit atop a New England radio powerhouse. Mississippi calmed himself as the classical piece ended, and he flicked on one of the turntables. To WBCN's loyal listeners, it was as if an alien mother ship had suddenly invaded their concert hall, panicking the startled dowagers out of their seats and scattering programs, purses, and fox stoles as patrons ran wildly for the exits. Hovering menacingly, Frank Zappa and the Mothers of Invention split the air with "Nasal Retentive Calliope Music," a bizarre and atonal assortment of spoken words, music, and farting sounds. Was Mississippi Harold Wilson some emissary from another planet, come to prepare the populace for a coming invasion? Not likely. But in those first few seconds of airtime, he provoked a change just about as jolting as an extraterrestrial takeover (at least one that was filmed in Hollywood). This initial shock led into his next choice, the breezy blues of Cream's "I Feel Free." With that first complete song, WBCN's new chapter, "The American Revolution," one that would span nearly forty-one years and five months, had officially stepped out of its mother ship.

As the prophet of "The American Revolution," Ray Riepen left a lasting impression with anyone who worked for him at WBCN or the Boston Tea Party. "[He was] an extremely intelligent man with a fair amount of W.C. Fields in him," Joe Rogers recalled. "When I met him he was living in an apartment in Cambridge with a mattress on the floor and a stack of books almost up to the ceiling. The man had one three-piece lawyer's suit and a couple of shirts. That was it. In the back of his Lincoln Continental was his laundry . . . in the trunk." Tommy Hadges described him as "a most unlikely entrepreneur" who drove his Lincoln "barefoot," also saying that his first meeting with Riepen "was in his beautiful, luxury apartment, but there wasn't a stitch of furniture in the entire place. To sit down, there was an orange crate! This was the guy that was going to take over a radio station?" Ten years Riepen's junior, future WBCN jock and program director Sam Kopper called him "our boss, forever in a pin-stripe blue suit; that could be daunting. But, he was a hippie in spirit, [if] not in dress or look." The "Master Blaster," cohost of Peter Wolf's eventual late night WBCN radio show, added, "Ray Riepen? He was a wild dude, man; he definitely had his own style. He had this big limo and he spent more time hanging out in that car going somewhere than he did in his house! He conceptualized the whole idea of the Tea Party and WBCN when people were just starting to question authority and be free." Truly a memorable character, Ray Riepen would shuttle in and out of Boston in barely six years, yet his

Ray Riepen, the hippie entrepreneur. Photo by Michael Dobo/Dobophoto.com.

tenure indelibly altered the city's cultural landscape, even if his name is often overlooked today.

Ray Riepen was a bright attorney who hit town from Kansas City to pursue a master's degree at Harvard Law School. By 1966, the seeds of the counterculture had been sown and were swiftly taking root. Change electrified the air, especially in America's college towns, where like-minded souls gathered from their diverse and staid homes across the country to collaborate and conspire freely on campus and in smoky coffeehouses, becoming part of some vast, liberal, petri dish. Riepen swiftly caught the buzz of the changing times firsthand in Cambridge and, as a voracious reader, soaked up the rich volumes of contemporary thought expressed by intellectuals all around him. There were opportunities out there for those who could visualize them, and even though he was a thirty-year-old graduate student in a scene that soon wouldn't trust anyone over that age, he still shared a great deal of the love-your-neighbor mentality that the hippie movement would emulate. "I've never done anything in my life for money," he explained. "I've done things antithetical to maximizing my money, because I've [always] wanted to do the most tasteful and innovative things." At the beginning of 1967, after he had clumsily and quite accidentally backed himself into a deal involving a failed South End coffeehouse on Berkeley Street called the

Moondial, Riepen's entrepreneurial spirit managed to turn that disaster into a launch of the city's eventual preeminent rock club, the Boston Tea Party. It was not without precedent. "I owned a jazz club back in Kansas City where Count Basie got started and John Coltrane played." (This was a measure of coolness not be lost on local jazz and R & B fanatic Peter Wolf.)

Soon the Tea Party was playing host to many of the city's hippest young bands like Bagatelle, the Lost, the Hallucinations (featuring Wolf), Beacon Street Union, and Ultimate Spinach. Riepen also began attracting smaller regional and national acts like Andy Warhol's Velvet Underground, Country Joe & the Fish, Canned Heat, Lothar and the Hand People, and Richie Havens. The legendary gigs that most people associate with the Tea Party—Led Zeppelin, the Jeff Beck Group, Fleetwood Mac, and The Who—were at least two years down the road at this point. But Riepen, as a lawyer and entrepreneur, was not all that keen on running the place: "I don't like to operate businesses. Once you get them and once you figure them out, it's not very elegant to be running them, so I hire people to run them." Steve Nelson, a Harvard-schooled lawyer who became disenchanted working in Washington (for NASA, no less, in the thick of the space race with the Russians), returned to Boston and went to see the Velvet Underground at the Tea Party to celebrate his twenty-sixth birthday. This was an extra special celebration night for Nelson: after being drafted years earlier and using his legal skills to avoid deployment, he had reached the magic age where Uncle Sam declassified him as eligible for service. "It was the end of May 1967; I went to that gig and I met Ray. A couple of months later, he said, 'I know you went to law school, you have a business sense and you know the music scene; would you be interested in becoming the manager of the Tea Party?' I thought, 'Yeah! That was so much cooler than working for the federal government!'"

So Steve Nelson went from launching moon rockets to moonlighting, taking the day-to-day management of the Tea Party off Riepen's shoulders. But things were perilous at best at the Berkeley Street location, and it seemed like Nelson's exciting new job might actually end up being the shortest one of his life. The Tea Party's bookings, although cool and hip to the underground scene, produced inconsistent results. While a pair of Country Joe & the Fish shows in August did terrific sell-out business, many other bands played to near-empty rooms. "It was a pretty small place," Nelson acknowledged, "especially when you think about where the music

THE BOSTON TEA PARTY
presents

JUNE 6 7 8   THE
GROUP IMAGE
THE CLOUD

lights by THE ROAD

$3.00   8:00 P.M.
53 berkeley street

JUNE 13·14·15

QUICKSILVER
MESSENGER
SERVICE

THE
HALLUCINATIONS

advance tickets:
headquarters east
george's folly
krackerjacks

THE
AMERICAN
REVOLUTION

WBCN·FM   104.1

stereo rock night and day seven days

WBCN presents concerts at the Boston Tea Party, June 1968, original poster.
Courtesy of the David Bieber Archives. Photo by Matt Dolloff.

business went after that." In a nightclub of this size there was little space for error. With a legal capacity of around seven hundred patrons, the Tea Party was not the kind of spot that could turn a huge profit unless the cover charge was jacked up, which Riepen refused to do. "I never made any money at the Tea Party," he said. "I was charging three dollars to get in [while] Bill Graham charged twelve, for the same acts!"

Don Law, who would eventually replace Steve Nelson as general manager of the Tea Party and go on to become Boston's most successful rock impresario, met Ray Riepen at Boston University (BU) where he was a student, as well as an instructor who conducted an educational workshop entitled "Evolution of the Blues." Law's early music education came from his quite famous father, Don Sr., who worked as a talent scout and producer for Brunswick and then Columbia Records, plying the American South for talent, which included the iconic Robert Johnson, Memphis Minnie, and even Gene Autry. Through his father, Don Jr. had developed some hefty connections on his own. He brought in blues artists to play and speak at his BU workshops, including the already-legendary Muddy Waters and his outstanding piano player Otis Spann. "[The workshop] got a lot of attention; the *New York Times* covered it," Law mentioned, "and that's how I met Riepen. At the time, he was struggling with the Tea Party; it was sort of hit or miss." The pair shared their backgrounds, and Law filled Riepen in on his recent successes as a young player in the music business. "I considered myself somewhat of an authority because I had worked as a college [booking] agent and had gotten Barry and the Remains their recording contract with Epic." Not only that, Law had also helped the Boston band land a spot on the road as an opening act for the Beatles. Riepen was impressed, quickly inducting Don Law into the Tea Party's inner circle. "It's been written that I was a lot smarter than I think I was," Law revealed, "or more Machiavellian. But the truth is that we were just kids who wanted to be in the music business. There wasn't any money in live entertainment anyway; it was all sort of a small business; we were doing it because we thought it was fun. Here it was: 1967 and '68, we really didn't have that much pressure; we were [just] having a blast. [But] then, the Earth moved!"

In the Tea Party's daily business conundrum, accounts receivable often didn't cover accounts payable, but the inner circle let it ride, counting on the next night's gate receipts to save the day. However, even as 1968 arrived with more consistent bookings, the profits were never outstanding. Plus, as Don Law remembered, his boss could be a financial problem himself: "Every time I'd get myself in a good position and start making money, [because] you had to have some around to make guarantees [with bands and their agents] and pay out bills, [Riepen] would just come in and take it all out to fund whatever else he was doing. I was always going, 'Oh no! I'm down to zero again!' I was always playing this game trying to keep the place afloat, [acting] as if I had resources . . . and I didn't!" Riepen's personal life also bordered on the dramatic, as Law recalled with a smile: "He was a really sweet guy, but he was just so difficult. I'd get a call at three in the morning . . . it was him: 'I'm over here in Belmont and I'm in jail and I don't have a license. Can you get me out of here?' He didn't have a license, and he would always drive without one. He had a Porsche one time and he was visiting some girlfriend on the Hill, then someone stole his Porsche. He said, 'Aw, fuck it,' went down the Hill and just bought himself a Volkswagen!"

Ray Riepen asserted that he didn't make any money off the Tea Party, but that's only true depending on how you do the math. To his credit, he kept the ticket prices low, insanely cheap by today's standards, and the hippies who were around then will always tell fond memories about how they saw Led Zeppelin or The Who for less than five bucks. Even so, Riepen did make money off the club, but he just spent it as quickly as he made it. Some of those expenditures, like financing his impending venture at WBCN and eventual stake in the *Cambridge Phoenix* newspaper, were clever and considered business endeavors. Others were more impulsive, as Don Law explained: "By 1969, 1970, there was this 'Free the Music' thing [in which] people thought there shouldn't be any charges for music. There was a lot of this, so we always tried to be sensitive about it and keep the prices down. Riepen took it upon himself to buy a Mercedes 600 limo! It was as ugly and corporate as you could get: a long black box that the only other people in the world used was the Secretary-General of the U.S. or the Pope. He'd pull up in this massive limousine at the Tea Party; I'd be out there trying to run things, and people are shouting, 'Free the music! Free the music!' He'd get out in his three-piece suit and stand there on the curb. I'd just go: 'Damnit!' And you know, if you looked in that limo, you'd notice that it was stacked with books. He would spend half his life in this thing in the back seat; it was really his apartment of sorts."

As a well-read intellectual, Ray Riepen noticed some recent developments in radio on the West Coast. In San Francisco, "underground radio" had made its debut on the FM band with a reasonable level of success. The idea began percolating in his head that the same could be accomplished in a place like Boston, with its profusion of freethinkers at the area's multitude of colleges and universities. "There were 84,000 students here and they were all starting to smoke dope," Riepen asserted. "They were obviously hipper than the assholes running broadcasting in America." He knew that a significant audience was hearing the music they desired every night at the

SPRING!

MAY

SUNDAY	MONDAY	TUESDAY	WEDNESDAY	THURSDAY	FRIDAY	SATURDAY
4	5	6	7 THE JEFF BECK GROUP & NICE	8	9 POCO & FAMILY	10
11 POCO & FAMILY	12	13	14 THE WHO *with Roland Kirk*	15	16 Joe Cocker & Grease Band *with Roland Kirk*	17
18	19	20	21	22 THE BONZO DOG BAND *with It's A Beautiful Day*	23	24
25	26	27 LED ZEPPELIN & zephyr	28	29	30 The Velvet Underground & The Allman Bros.	31

JUNE

1	2	3	4	5	6 Delaney & Bonnie & Friends & *the serfs*	7

The Boston Tea Party 338-7026   WBCN 104.1fm

The Boston Tea Party schedule, May/June 1969. Courtesy of the
David Bieber Archives. Photo of schedule by Matt Dolloff.

Tea Party, but they were not hearing that music on the radio. Up to that point, Boston was ruled by the format that had been dominant in America for over a decade: Top 40, with its tight playlist and hyperactive teams of shouting DJs to introduce the songs. The recipe of this winning formula was based on repetition. Since listeners remained locked on a radio station for discrete periods in the day based on their own personal schedules and preferences, it was important that while tuned in, those listeners heard the songs they absolutely loved. Relentlessly rotated on the air, that list of "40" selections ensured that the most popular songs would be heard by the highest-sized audience possible. A station programmed successfully in this manner could generate huge numbers of listeners throughout semi-annual ratings periods and, as a result, demonstrate to potential clients that it was the one to purchase advertising on, then command top dollar for every commercial sold.

Top 40 radio stations in the sixties weren't just marked by their choice of music; there was also a style or aesthetic behind the sounds of those frequencies. Program directors encouraged their DJs to shout with abandon into the microphone, modulate their voices up and down in exaggerated or phony exuberance, and deliver high-powered raps with machine-gun velocity. Most of what an announcer needed to say in a break between two songs could be accomplished by talking over the fading music of the first and the instrumental introduction to the second, before the singer's voice kicked in. Radio station engineers wired up echo devices so that DJs could project their voices with the booming, amplified voice of God (or at least Charlton Heston as Moses). Inane collections of sound effects, bursts of fast-paced phrases, and clips of words snipped from comedy records were produced into short station IDs (call letters plus city of origin) to link up songs and constantly remind listeners what radio station they were listening to. DJs had their own theme music, jingles, and personalized breakers to instantly convey their identity to the audience. "Cousin Brucie" (Bruce Morrow) at WABC in New York City assembled a brief, instantly recognizable montage of voices yelling out his name, while closer to home at Boston's WMEX-AM, celebrated announcer Arnie "Woo-Woo" Ginsburg earned his nickname by punctuating on-air raps with blasts from a train whistle.

As Lulu's "To Sir With Love," "The Letter" by the Boxtops, "Windy" from the Association, Bobby Gentry's spooky "Ode to Billie Joe," and "Groovin'" from the Rascals became the biggest AM radio hits of 1967, immense changes were afoot in the youth culture and, eventually, in the media that served it. Just as the Beatles had quickly outgrown playing neat two-and-a-half-minute pop masterpieces by absorbing fresh influences and exploring new adventures, fans began searching for more as well. Young, unsullied minds lay open to innovative approaches in art, music, spirituality, lifestyle, health, sex, and politics. The sixties became the stage upon which these great changes occurred, revolutions approaching and flying by like road signs on a highway. The specter of the Vietnam War haunted the country, uniting America's teenagers and propelling them forward with urgency. Why would an eighteen-year-old seeker wait to experience an LSD trip or pass up a weekend retreat with some Indian guru, when he knew that the following week he might be plucked by his draft board and sent off to dodge shrapnel in Da Nang. This great injustice, visited upon the youth by political leaders seemingly ensconced on Mount Olympus, was an accelerant poured on the fire of the times.

Encyclopedia Britannica summarized this turbulent time in American history: "The 1960s were marked by the greatest changes in morals and manners since the 1920s. Young people, college students in particular, rebelled against what they viewed as the repressed, conformist society of their parents. A 'counterculture' sprang up that legitimized radical standards of taste and behavior in the arts as well as in life." A potent antiwar movement poured out of the American campuses and took to the streets. Swiftly rising opposition to the U.S. Air Force bombing of North Vietnam brought a hundred thousand demonstrators into New York City in April 1967, and an October march on the Pentagon drew almost the same number. The black population, still largely unable to claim the rights won during the Civil War a hundred years earlier and guaranteed under the Constitution, had waited long enough. Simmering anger boiled over into violent action, as evidenced by race riots in dozens of American cities, particularly Watts in L.A. and other revolts in Newark and Detroit. In the latter, rampant looting and one thousand fires destroyed almost seven hundred buildings in less than a week. The nonviolent civil rights movement, led by Martin Luther King Jr., encountered stiff resistance at every turn but steadily grew, its ranks swelled by sympathetic supporters in the antiwar movement. Many young protestors also rejected the capitalist stance of their parents, instead taking it one day at a time to explore alternative lifestyles. The recreational use of marijuana and LSD soared while drug-culture gurus like Timothy Leary urged his followers to "turn on, tune in, and drop out."

It's no mystery, then, that music and the radio stations that played it were due for a rocket ride of their own. History points a big finger at the Beatles' June 1967 release of *Sgt. Pepper's Lonely Hearts Club Band* as the flash point. There were others, and some came before, but the Beatles had the attention of the world as pop music's number 1 spokesmen, so the group's latest work was greeted with far more attention than any other. Paul Evans, writing in the *Rolling Stone Album Guide*, lauded *Sgt. Pepper's* as no less than "the most astonishing single record of popular music ever released." Never before had a band with this magnitude made such a bold artistic statement, shrugging off its monochrome past and embracing a fresh Technicolor one. The group realized this, evidenced by the cover photo featuring the waxen Fab Four replicas in charcoal-grey suits standing solemnly at their own funeral, in front of the "new" John, Paul, George, and Ringo in colorful marching-band outfits. The Beatles didn't necessarily need the promotion of a hit song, so the band chose not to release any singles from *Sgt. Pepper's*, instead presenting the twelve-inch vinyl record as a whole piece not unlike a complete classical work of old. Top 40 DJs who didn't want to be left out of the newest Beatles project blew the dust off their turntable speed levers and moved them (maybe for the first time) from 45 rpm down to 33, and then started playing songs off the album. *Sgt. Pepper's* became the largest-selling album in America from July into October 1967, ushering in an era of sophistication in popular music that hadn't existed much before. Now, many listeners began grooving to complete albums rather than stacking up a pile of three-minute singles on their record player.

Tom Donahue, a successful West Coast Top 40 disc jockey and businessman, had read the same tea leaves as Riepen, but his vision came a year earlier. He proposed a radio show embracing some revolutionary elements: mixing album tracks, creating long sets of music drawn from many genres of music, speaking conversationally, and avoiding the wisecracking gimmicks of Top 40 delivery. He unveiled his ideas to the management at KMPX-FM in San Francisco, which featured ethnic programming at the time. Facing financial challenges, the owners decided they had nothing to lose and allowed Donahue to experiment with a free-form evening show based on folk, blues, and rock music. The new approach became an immediate sensation, especially amongst Bay Area students, and within a few months KMPX adopted the revolutionary format full time. Since that complete changeover occurred on 6 August 1967, Tom Donahue is credited with having started the very first "underground radio" station in America. Soon, the owners called on Donahue to repeat his success by extending the same type of programming to KMPX's sister station in the Los Angeles area, KPPC-FM in Pasadena, which had previously been broadcasting classical music out of a church basement to no great financial reward. With underground radio visualized and the prototypes already flying on the West Coast, why couldn't the same also work in Boston?

Riepen considered the FM radio band as virgin territory for his experiment and attended a 1967 FM broadcasting conclave in Washington, D.C., to share his views and garner feedback. The reaction, however, was universally negative, even condescending. "I told them what I wanted to do and they asked, 'What's your background?' I said 'I'm an attorney, I've got a business degree; I'm in the concert business.' They said, 'Well, you better go back and practice law because you're the dumbest son of a bitch we've ever heard of! You can't have breaks with no talk-overs and no jingles . . . or [the DJs] using conversational tones and playing eight to ten minute [musical] pieces!" Ridiculed at the gathering, he assumed an "I'll show you!" attitude and forged on even more intently. With Steve Nelson and then Don Law taking care of the Tea Party, Riepen began examining Boston's FM band in great detail for a likely spot to base his experiment.

Riepen's research led him to 104.1, WBCN-FM, a struggling classical music station that had sparked up its transmitter nearly ten years earlier on April 24, 1958. The owner and president, Theodore Mitchell Hastings, possessed a colorful and distinguished past as one of the very pioneers of FM radio. With a genius-level intellect, the man could easily grasp technical concepts that most found impenetrable; he was a true eccentric, with an almost childlike view of science, who commanded a great deal of respect from the engineering community. Like Ray Riepen, Mitch Hastings went to Harvard, graduating in 1933, and then formed an electronic research lab named General Communications, which performed extensive work in sonar and signaling technology for the navy during World War II. By 1951, he had become fascinated with the commercial possibilities of the still largely untouched FM radio band, with its low interference and static, as well as stereo feasibility. At the time, there were less than two dozen FM programmers in the entire country, with growth in new frontier virtually nonexistent. Hastings recognized that the newer radio band had to be readily accessible to the masses before its technology could expand on a grand scale. The key, he felt, was to create an FM radio for the car, which he developed with Raytheon engineer Ed Brooks and marketed to listeners on the handful of FM stations around the country. Hastings became somewhat famous for this, and after a few further developments and inventions (including a pocket FM transistor radio) he decided to jump into the world of radio station ownership himself.

Pouring his own funds into the project and obtaining additional money from several financial backers, Hastings formed the General Broadcasting Corporation, later to be known as Concert Network Inc. The company expanded rapidly, acquiring a handful of powerful FM radio stations up and down the East Coast. According to Ron Della Chiesa, "His dream was to create a network of stations that would program classical music all over the country on FM. Up to that point, classical music was on AM; you couldn't get the full frequency of it. [But] when FM came in, it was like a third dimension; you really got the depth and sound." Hastings acquired and then changed the call letters of each station, putting a "CN" in each to designate Concert Network. "So, beginning his dream were these stations:

Mitch Hastings struggled to keep classical music WBCN afloat.
Photo by Sam Kopper.

WNCN in New York City, WHCN in Hartford, WXCN in Providence and [for a time] he had one on Mount Washington, WMTW." Added to this list was an additional property in Riverside, Long Island, renamed WRCN and, of course, WBCN in Boston.

WBCN began broadcasting classical music on T. Mitchell Hastings's birthday in 1958; and so his dream took flight. The idea was, as Hastings told Alan Wolmark twenty years later in *Record World Magazine*, "to go forward and develop FM broadcasting into the great public service it should be." Hastings organized his network of stations into what he termed the "Golden Chain." In effect, he could nearly pull off systemwide live broadcasts of the Boston Symphony and other classical programming by originating the transmissions on WBCN, which would then be picked up by high-gain antennas at 'XCN in Providence and rebroadcast to its own audience. In turn, that transmission would be relayed further down the chain to the stations in Long Island, Hartford, and New York. Although Boston was the originator in most cases, programs could also be sent back up the network in the reverse direction. In an age before satellites blanketed the skies and routinely linked broadcast stations around the globe, T. Mitchell Hastings was able to bypass the high costs of using phone lines to achieve this same purpose.

T. Mitchell Hastings possessed another side to his personality, quirks and oddities that often challenged those around him. "[He] was a whack job, and I don't mean that meanly; but he was just a very strange guy," future WBCN program director Sam Kopper pointed out. "In addition to his sort of visionary eccentricities, he was a religious and spiritual seeker." Not secretive about his beliefs in the supernatural, spiritual, and paranormal realms, Hastings and his wife Margot were regular acquaintances of Edgar Cayce, perhaps the most famous clairvoyant of the time. The couple sat with the seer on a number of occasions to seek advice and gain glimpses into a future that Cayce would reveal during his legendary and well-documented trances. Some of these predictions formed the basis of Hastings's critical business decisions. "[He] was a wildly eccentric guy," Chiesa agreed; "he believed in Atlantis, Cayce and mystics. [Hastings] lived in another world; he could have been an extra-terrestrial. But, he also believed in the power of classical music to awaken the spirit and the mind. He was a purist; there was a depth to what he wanted."

Ray Riepen, who would soon lobby to introduce rock music to WBCN, understood that Hastings was deeply committed to programming his personal love of classical music on the station. But he could also see the weaknesses in the owner's thinking that jeopardized what he had built. "[Hastings] was a visionary guy in early FM who had put this little network together," Riepen said, "but he was not a businessman; nor was he anybody of any taste or discernment." In his own role as a program director and classical music expert, Chiesa had to agree that even though Hastings clearly loved the music, his tastes were quite finite. "He knew what he liked, [but] he didn't really know that much about classical music. He would call the station occasionally and yell, 'Get that off the air! I'm not enjoying that'; and it would be in the middle of something great like 'Scheherazade' or even Beethoven. He could call and disrupt the whole flow of what you were doing, and you had to pacify him by saying that you would do it, whether you did or not."

Personal peculiarities aside, Hastings was still quite a man to respect, but his innovations and his achievements could only take him so far. It soon became obvious that the classical music he loved so dearly would not generate the amount of advertiser interest needed to keep his stations thriving, or even financially afloat. Chiesa remembered, "There were times where we didn't get paid for three or four weeks." The bill for the UPI (United Press International) newswire service went overdue for so long that the company sent workers to disconnect the teletype machine and haul it away. This did not end news reporting on WBCN, though: "Hastings told us, 'Just read from the newspaper'; so that's what we did. He also had problems with his rent [at 171 Newbury Street]. The building was owned by an old architect named Edward T. P. Graham, who was becoming quite feeble and would just sit around in the office below the studio with an elderly secretary. I heard T. Mitchell talking to the secretary once when he was several weeks behind in rent, saying that he'd like to do a memorial program for Mr. Graham when he passed away. So, he was willing to trade air time for rent!" Plus, he had proposed the deal for a sponsored radio eulogy before the subject had even died! Sometimes Hastings took the elevator down to the street and personally canvassed the crowds passing by on the sidewalk, looking for people who would donate money to the station. Time and time again, though, classical music lovers or friends of Hastings would arrive in the nick of time, like the 7th Cavalry, to put up fresh funds to cover the bills.

As Hastings's financial woes deepened, he became increasingly desper-

Ron Della Chiesa, WBCN's classical music impresario (from 1987).
Photo by Dan Beach.

ate and open to almost any new possibilities for cash flow. In the early sixties, radio programmer Marlin Taylor had developed a winning format called beautiful music or easy listening. Taking advantage of the advanced fidelity available on the FM band, Taylor blended innocuous vocal songs with light orchestral hits to produce an inconspicuous mix for background listening. By 1966, after observing the success of several stations with this new format, Hastings contacted Taylor. Soon the sounds of Montavani, 101 Strings, Johnny Mathis, and the Ray Coniff Singers wafted with saccharine sweetness out of WBCN's transmitter. "There was a period of several months where that's all that we played," Chiesa recalled. "Mitch liked beautiful music probably more than classical, because it was so bland and more accessible to him. But it didn't work; the format petered out and we went back to the classical." Chiesa also revealed that the first song played in the new format was "The Ballad of the Green Berets" by SSGT. Barry Sadler. Considering what WBCN would eventually become, the presence of beautiful music and particularly this song, a promilitary and Vietnam War anthem that went to number 1 in America and stayed there for five weeks in 1966, remains remarkably ironic.

The "Golden Chain" that Hastings had worked so hard to forge began to fall apart: WRCN-FM on Long Island was sold, WXCN-FM in Providence cut loose in October 1963, and WNCN-FM in New York unloaded the following year. The network had been whittled down to just WHCN-FM Hartford and WBCN itself, and Hastings was on the brink of losing even these. Ray Riepen observed, "[WBCN] was in Chapter II bankruptcy; they were not making their payments and it was going to go down." Frantic, the board of directors granted Riepen an audience to present his views about the merits of a free-form, rock music format. Joe Rogers related, "Ray went to them and said, 'You have absolutely no income whatsoever during the overnight hours. I can provide you with actual listeners who might, in turn, generate sponsors who, in turn, could bring revenue to your station. Face it, you have nothing to lose.'" Don Law added, "Mitch Hastings was such a classical music lover and saw FM as the salvation for that music." He shook his head and marveled: "Riepen actually talked him into changing his format." It wasn't easy, though, as Riepen picks up the story: "Mitch Hastings was appalled at this thing and fought it all the way. The board, though, were businessmen; they were old friends of his that were trying to save him. The classical music was continuing to flounder and they understood the deal. They were so desperate, they gave me [the time slot] after midnight." Actually it was 10:00 p.m. to 5:00 a.m. and strictly on a trial basis, but it was a shot. Ray Riepen congratulated himself, got on the phone, and gave the go signal to Joe Rogers.

There was an immense freedom; you were defined by your
personality and your musical tastes. That was very intoxicating and
exhilarating for everyone involved because you realized you were
part of something incredibly new. PETER WOLF

A RADIO COMMUNE

Ray Riepen gathered his recruits for the new WBCN air staff, but, in a surprising move, avoided chasing after professional disc jockeys. "I didn't want people who were in radio trying to figure out what we ought to do, because they couldn't. They were swimming in the sea of a Top 40 world; all over-hyped and screaming. As far as [producing] something tasteful or smart, they didn't have that kind of vision." Joe Rogers marveled at the idea: "He could have staffed the station with real announcers, but he stuck with rank amateurs because that's how he saw it. In the end I guess it was the right decision, but it was a peculiar thing to do at the time." The place where Riepen could find the type of people he wanted, in an environment that embraced freethinking and was uncluttered by format and focused on the music, was on the Boston area's many student-run radio stations. With Top 40 dominating the AM band and classical programming holding down the FM side, college radio was the place where the sounds of the burgeoning folk and blues revival and growing rock revolution could be heard.

"When Ray originally decided to start a radio station, he went to the MIT and Harvard stations: WTBS and WHRB, to find people who would be willing to do this sort of alternative [he envisioned], and that was the core that started [WBCN]," Tommy Hadges pointed out. Both Hadges and Rogers were Tufts University students but had found their way onto WTBS. Riepen checked them out as well as several other jocks at the station including Jack Bernstein, plus Steve Magnell at Tufts and Tom Gamache on Boston University's WBUR. He passed on Gamache but approached the others and set up a meeting at his nearly unfurnished Cambridge apartment to present his plan to them. After floating the idea around the room and receiving assurances from the jocks that they were certainly interested, not much else was discussed. Even later, when Riepen had confirmed that WBCN was giving the jocks a shot, there was virtually no planning or guidance from their mentor. Rogers elaborated, "No serious preparation. We knew there was a date coming up and whatever that was [in his best Riepen imitation], 'Be ready and go do your god-damned radio show! Don't fuck it up!' There was no discussion of the format; it was, 'Just do what you've been doing on these college stations, but do it better!'" Riepen added, "They didn't have the concept that I had, but once we set the parameters, I didn't restrict them. I wasn't going to program the station; the idea of it was to be free-form and spontaneous."

It was only three weeks after their initial meeting that Rogers got the call to get down to 171 Newbury, but in that time Riepen had also pitched his idea to a familiar face at the Tea Party. "Peter Wolf sang in a group called the Hallucinations, who were mostly guys from the Museum Art School; I used to book them from time to time. Peter Wolf was kind of a star; he had the moves. He was a smart guy and very together in certain ways." Showmanship, a keen ear for what the people wanted, and an encyclopedic knowledge of R & B and blues combined to make this gifted neophyte an excellent candidate for some musical vocation, in spite of his chosen field of study. As Wolf put it in the 2003 essay compilation *Martin Scorsese Presents the Blues*, "I was a student looking to become a painter somewhat in the German Expressionist manner. But then I was sidetracked by the blues." Born and raised in New York City, Wolf's entire life was surrounded by the arts: learning the basics of drawing, filling canvas after canvas, and then seeing jazz greats like John Coltrane and Miles Davis in the Village and soul giants James Brown and Ray Charles at the legendary Apollo. Upon arriving in Boston on his thumb, Wolf pursued his love of music with a passion, settling in Cambridge just a stone's throw from Club 47, the most important early sixties terminus for local and touring folk artists in New England. Soon the singing novice and harmonica trainee became a regular at Club 47 and Boston's Jazz Workshop, meeting and jamming with A-list favorites like Muddy Waters, John Lee Hooker, Howlin' Wolf, and James Cotton. Peter Wolf's apartment morphed into a haven of sorts for bands playing Club 47. Waters and his band became regular visitors, relaxing, playing cards, and cooking dinners before and after their sets while their scrawny host built some solid friendships and soaked it all in.

Wolf threw his musical passions into the Hallucinations and began to perform on a steady basis around the area, finding a regular home at the Tea Party in early 1967. The word started getting around about the skinny kid who always dressed in black and sang like he grew up on the south side of Chicago. Riepen saw not only a talented lead singer for a band here, but one who could equally mesmerize on the radio. He piqued Wolf's interest with his plan for WBCN. "[Riepen] was in my apartment one day and he was all full of steam and pumped up, telling me about this business venture he was getting into," remembered Wolf. "He was going to take over this radio station, but he could only come in slowly by doing the evening [shifts], and if I had ten thousand dollars I could partner with him.

'Ten thousand dollars? Ray, I don't have *ten* dollars.'

'Well listen, you've got ten thousand records! How about [being] a DJ and helping me get together a staff and build up a library?' And I loved radio; when I was growing up, it was so important to my knowledge of music. The idea of doing a radio show was exciting."

"He had a beautiful understanding of what to do with the radio," Joe Rogers said of Wolf. "He understood something about show business and how to grab people's attention, then what to do with them once you had them." Charles Daniels, the "Master Blaster," added, "We used to hang out in [Harvard] Square and spent a lot of time at Peter's place because that's where it all seemed to happen; it was like a big family. He always had the best records."

Rogers would be first, stepping across the threshold at 10:00 p.m. and into a new world not unlike Neil Armstrong's small step a little more than a year later. Wolf would then arrive for the overnight shift, often, as demonstrated in future months, literally out of breath after rushing over from a gig. Indeed, it happened the very night he made his WBCN debut; the Hallucinations performed at the Boston Tea Party with the Beacon Street Union. WBCN would eventually broadcast from a back room at that dance-hall, and many have speculated that it was a simple matter on 15 March 1968 for Wolf to run offstage, towel himself dry, and jump in behind the turntables. But Rogers is quick to refute this. "Many people think that the first [radio] show was at the Tea Party; I've given up trying to correct them. We began where the proper studios were located: 171 Newbury Street."

"He's absolutely correct," Tommy Hadges concurred. "I remember [Joe's] hand shaking so much I might have been the one running the turntable!"

Don Law also confirmed that the debut occurred in the WBCN studio on Newbury Street. "The very first time the needle dropped on that first record, I remember being in the next room with Riepen, [who was] pop-ping pills and drinking champagne, looking through the glass. Riepen was readably nervous as the calls and the complaints started coming in." Although WBCN's previous format had not kept the station in the black, the few classical music listeners who were tuned in at that late hour erupted with dismay, lighting up the phone lines in disbelief. But calls of support from a new audience were just as vocal and would increase dramatically over the next few weeks. "God bless Peter Wolf for many things," Rogers stated, "but among them, he was the only one who had the presence of mind to instruct listeners that it would be a good idea to send in letters and postcards if they liked what they heard. An awful lot came in too; I saw some 30-gallon barrels of letters. It made a significant impression on the people [in the office] downstairs." This was important since, as Rogers added, "There were objections to us; we weren't entirely welcome at 171 Newbury Street. We weren't . . . uh . . . what they were used to at a classical radio station."

"They were from the Nixon conservative era; but the [new] staff was anti-Vietnam, progressive left-wing," Wolf recalled. "Sandra [Newsom] was this woman that ran the old 'BCN office, sort of like a gal Friday: did all the billing and booking and stuff. She had this huge poster of Barry Goldwater on the wall. Goldwater was Attila the Hun to us! She couldn't stand these dirty hippies; didn't want them in the office, even though they were creating this thing."

Peter Wolf in 1968. "He had a beautiful
understanding of what to do with the radio."
Photo by Charles Daniels.

"I guess they were afraid we were on acid all the time and we'd steal the records or something," Riepen chuckled.

Wolf didn't just arrive at the station with a box or two of R & B records to play; he arrived with a plan. He had a theme song to start the show: "Mosaic" by Art Blakey and the Jazz Messengers, over which he'd rap introductions for a cast of characters, both real and imagined. He came with an entourage, including neighbor and friend Charles Daniels, who was becoming a celebrity as the Tea Party's house emcee. Daniels recalled, "The radio was just a reflection of what we were doing at that time. I'd talk about the Tea Party, who I'd hung out with, or if I was wearing some funky new underwear. I'd take my pants off [in the studio] and say, 'check this out!'" Daniels evolved into a sort of cohost: an Ed McMahon to Wolf's Johnny Carson. "I started calling him the 'Woofa-Goofa' and he started calling me 'The Master Blaster.' We'd joke back and forth, than jump right into a song," Daniels explained. "It wasn't comedy radio, but a lot of slapstick and jiving, with really, *really* good music." One of their favorite bits was an old slice of black folklore called "doing the dozens," as represented by "Say Man," one of Wolf's favorite songs by blues great Bo Diddley. The two would launch into a lengthy and friendly rant based on the "your baby is so ugly" motif that Diddley and colead singer Jerome Green jived over on their 1958 Chess single. Wolf and Master Blaster volleyed, piling on the insults until one or the other, both usually, collapsed in laughter. "It was very funny," Rogers laughed. "I was into blues, but R & B was [Wolf's] specialty. It was new material to me, and I'd say, 98% of our listeners, who would have heard nothing like it."

"My show was an enigma. Because of my relationship with Ray, I was given carte blanche," Wolf explained. "I brought all my records and 45's in; I'd program Billy Stewart, lots of blues, R & B, rock and roll, Van Morrison, deep Bob Dylan tracks, Tim Hardin and a lot of singer/songwriters like Jesse Winchester. I'd play things that were not in the main vein of what 'BCN was going after, which was Country Joe and the Fish, Jethro Tull and the new music that was coming out." In his husky growl, he'd grant blessings on the "stacks of wax," "mounds of sound," and the "platters that matter." But Wolf still had to learn the basics of radio, as he related:

> I got a call one night from Ron Della Chiesa. In his great, great voice, he said, "Hey man, what's wrong with you? Where's the ID?"
>   And I went, "What?"
>   "The ID!"
>   "What's that?"
>   "The station identification!"
>   "Well, what's that?"
>   "Don't you know you've got to give an ID every half hour and hour?"
>   "Oh!" So, then I said "WBCN" on the air when I was supposed to. He called back and said, "No, no! You've got to name the city too; you have to say the whole thing: WBCN, Boston!" [Wolf laughed.] So, you know, he taught me to do that!

Within a week of his first show on 'BCN, Peter Wolf ran into Jim Parry, a Princeton graduate and self-professed "unregenerate folkie" who had made his way north to the folk music Mecca of Harvard Square. "I was working at Club 47 and I knew Wolf from the music scene," Parry remembered. "He said, 'I'm doing this show at this new station,' which I had vaguely heard about, 'but I've never engineered before and I have no idea how to do that.' I said, 'Oh, gee, I did that when I was in school. I'll do it for you.'" So, as simply as that, Parry joined the cast of characters, manning the controls, pushing the buttons, and cueing up records for his new boss. "[Wolf] was always a total music-head; he had a shit-load of very esoteric records." As dawn closed in and began to dissipate the nocturnal spell that Wolf had cast during the wee hours, he would launch into a closing rap in which he thanked his cohorts and bid adieu to another long night. "Jim Parry—looking so merry" would get his nod as well as "Master Blaster" and "the Kid from Alabama keeping it all hid" (who was Ed Hood, a friend from the Square who studied at Harvard and had been the star of Andy Warhol's 1965 film *My Hustler*). Wolf would also mention the station engineer he dubbed "Sassy John Ten-Thumbs." Jim Parry explained, "He was an MIT dropout and sort of lived in the back room of the Newbury Street studio. There were cubicles back there and he walled off this area with science-fiction books. He was essentially homeless and lived there for about a year."

Also working the board for Wolf on occasion was Al Perry, who had been at WBCN for six months at that point. "Ron Della Chiesa actually hired me when [the station] was still classical. I swept floors; then I got my FCC license and did a little news and weather at the top of the hour. When Ray Riepen came in, I started selling [the station]." This meant that he was working the streets for the sales department by day and jamming on the air most of the night. "T. Mitchell Hastings had actually built a shower and a bathroom downstairs with a roll-out couch," Perry recalled. "I'd get off at six in the morning, sleep for a while, shower and shave, go to a sales meeting at 8:30, then hit the road. So, yeah, I was pretty tired. Wolf would yell at me because I'd fall asleep and the record would be spinning around . . . 'Wake up, you sonuvabitch!'"

"I'd come in and Al would be lying there on the floor," Jim Parry laughed. "He slept with his eyes half open for some reason, so we were never sure if he was asleep or awake." Master Blaster added, "If I like you, at some point I'm going to give you a name; that's just how I am. So, I started calling [him] Crazy Al, because he was *always* doing something at the station; he never got to go home!"

Shortly after the initial shows at the Newbury Street studio, the broadcasts from Riepen's nightclub, the Boston Tea Party, began. "Ray spent the money for a Sparta [control board] and that was put over at the club," Rogers explained. "Then it became normal for us to broadcast from there on weekends."

"I got the board for two or three hundred dollars and put it in the dressing room, and of course, it didn't hurt us if we could talk to people like The Who, Led Zeppelin, Jeff Beck, or Rod Stewart between sets!"

"There wasn't really a backstage dressing room," Tea Party general manager Steve Nelson explained about the layout of the club. "It was originally built as a church, so there was only this one little room, almost like a ready room, where you could go and sit, then jump up on stage. But the big room off of my office on the other side of the building we called the 'back room'; that was a hangout space and dressing room. Bands would walk right from that room, through the crowd, to the stage, and vice versa. There wasn't really any privacy; if you were in that room, then you got to hang out with the band."

"I remember music coming through the door while we were on the air," Jim Parry recalled. "Soundproofing? What's that?"

"We'd put a blanket over our heads so we couldn't hear the concert through the microphone when we were trying to talk!" Wolf added. Jim Parry also shivered at the memory of the Tea Party, because after the concerts were over, the staff would clean up the mess, flick off the heat, and leave. "The place was locked down after hours. It was still, basically, winter. [Wolf] was doing all night, and it was cold as hell! I was in a pea coat and gloves, and he was wearing some kind of winter coat for about the first month. He periodically put on a long cut, and we'd break into the ballroom and steal sodas from the concession stand there. He practiced screaming on the stage, [doing] his Howlin' Wolf imitation."

Ray Riepen's strategy was dependent on showing Mitch Hastings and wbcn's board of directors some significant sales progress during the experimental hours of free-form rock. "I told [the sales department], 'Listen, we don't have any [ratings] numbers. I don't want you to go into the big agencies and talk to these guys that want to see our numbers. I want you to go into small stores and sell them. I want you to bring in money . . . five dollars at a time." But, word-of-mouth awareness of Boston's latest radio format spread at an astonishing pace. When Al Perry and his colleagues hit the streets, they found their potential customers already well aware of

"Jim Parry, looking so merry." Photo by Dan Beach.

what was going on at 104.1. "Back in those days," Perry related, "the head shops pretty much stretched from Massachusetts Avenue and Boylston Street in Boston to Cambridge. Typically, you'd go in and say, 'I'd like to talk about radio; I work for WBCN.'

'You do? Who are you?'

'Al Perry.'

'Al Perry? I was getting laid last night listening to you!' Then they'd start signing up."

Soon the money began trickling in as the sales team mined their fresh vein of paisley- and Nehru-clad clients. These new buyers actually preferred the late night commercials being sold during "The American Revolution," since most of their customers followed a far more nocturnal lifestyle than the nine to five workers that Mitch Hastings's team had pursued. Sam Kopper, newly returned to Boston after graduating from Syracuse University, briefly took a job selling for Riepen alongside Al Perry and longtime veteran Jack Kearney, who had also been at BCN during the classical days. "Selling [time on] the station was not hard," Kopper revealed. "[WBCN] was coming out of every window in Back Bay, so places like head shops and record stores totally got it. By the way, it cost eleven dollars for a 60-second spot!" Though it was easy selling time, Kopper despised doing it, lasting only two days in the department. Fortunately, he got a second shot as a DJ one night in May, filling in for Joe Rogers, who was still in school and had forgotten he had to take an exam. Kopper's timing couldn't have been more perfect. Within a matter of days, Hastings announced that Riepen had proven his point, and 'BCN went on the hunt for full-time jocks to fill out the schedule. Except for some weekly recorded religious programs, "The American Revolution" had gone twenty-four-seven. Rock and roll was here to stay!

"Traditionally, it's been difficult to sell FM time during prime periods, let alone after 10 o'clock at night through 5 in the morning," explained Riepen to the *Boston Sunday Globe* in early 1969. "We showed Mr. Hastings we could sell 'spots' at those ungodly hours, so he figured we should sell even more during prime time. Due to the tremendous community response we went 24 hours a day as of last May 20th [other sources indicate May 17th] with the rock or contemporary music format, and the financial feasibility of such programming has been proven."

"We had these pictures of opera singers and conductors on the wall at the old 171 Newbury Street studios, and I'll never forget taking them all down," Ron Della Chiesa recollected. "It was the end of an era, and I was tinged with sadness [at] the fact that it couldn't work." As Chiesa moved out, Riepen moved his people in. Certainly Joe Rogers, now well known around town as Mississippi Harold Wilson, would remain along with Peter Wolf, whom Rogers labeled "the key to the station," in a 1972 story in the *Real Paper*. Sam Kopper, with his experience hosting a folk music show on WAER-FM in Syracuse, actually had more radio experience than any of the others and was tapped to do mornings. That slot, as Kopper related, was not coveted: "It was considered lower than even overnights because hippies and college students didn't get up that early. I developed a completely different philosophy: whereas most morning shows were about yelling or bells and whistles to wake you up, my idea was to play energized music. I would take people through these changes: a set might begin very gently, as mellow as Nick Drake, but then work its way up to higher and higher energy. One of my favorite compliments was from the first chair violinist of the Boston Symphony who wrote me a letter flipping out over a segue I'd done from Franz Schubert into BB King."

Tommy Hadges, Al Perry, and Jim Parry all earned shifts, cashing in on the experience they'd received helping Wolf and Rogers on their initial shows. But Riepen still needed more talent and, in an inspired move, imported underground radio veteran Steve Segal, who had been working for the legendary Tom Donahue at KPPC-FM in Pasadena. Segal flew the coop to arrive in Boston in June 1968. "I remember my first night, landing at Logan Airport; Riepen was waiting for me. He meets me where the passengers and guests were mingling; we shake hands and he says, 'My Cadillac broke down on the way here and I had to pull it off the road, so we're going to have to walk back to Cambridge. Don't worry, it's not too far.'" Segal laughed at the memory of Riepen's blatant fib. "I swear to God, I walked all the way from the fuckin' airport to Cambridge! It took me forever, and all I can remember was him telling me about Peter Wolf, Mississippi, and all these guys. His basic take on them was that they were all nuts; but, of course, he was the craziest of them all."

Riepen, crazy or not, knew what he wanted, promptly putting Segal on the air and installing him as the temporary head of programming, although his responsibility was not as official as a modern program director's. Since every jock at WBCN was encouraged to play their own musical blend, Segal's taste and greater experience would merely serve as a beacon for the others to follow. "He had been mentored by Tom Donahue and that whole scene out there," Sam Kopper mentioned. "Whereas we had perfectly good makings for what would become a valid underground rock station, bringing in Steven at that time made a huge difference. He was very much the guru for our station."

"It just came naturally to him," Al Perry added. "He'd play three songs and you were just dying to hear what he'd have to say."

"It was 1968. We were all socio-political; we were all involved," Kopper pointed out. "The commitment to [that] was almost as important as the music, but Steven did it better. I consider him, maybe, the Howard Stern for NPR [or] a left-wing Rush Limbaugh, other than the fact that he made it happen with music." Charles Laquidara, who was not yet in the picture at WBCN, would summarize years later, "Steven was the most brilliant disc jockey that ever existed; Howard Stern paled next to him."

"Everyone treated me with this great respect," Segal explained, "but as far as me being the big radio veteran from the West Coast, I'd only been on the air for six months in L.A. by that point! I was excited to be [at BCN]; I met [everyone] and they all seemed to be, not professional radio guys, but

The original WBCN jocks out on the town: (from left) sports reporter Bud Collins, Steve Segal, Peter Yarrow of Peter, Paul, and Mary, Joe Rogers, Sam Kopper, and Al Perry. Courtesy of the Sam Kopper Archives.

funny, wacky . . . out of their minds. I felt like I was home." Segal noticed a fundamental difference between the underground radio of the East Coast as opposed to his former home.

> The West was mostly ex-star disc jockeys who had made a revolutionary change: they realized that albums were going to kill 45s and they better be where the action was. [They were] super-professional guys re-creating radio with plenty of thought about the things they wanted to present. But in Boston we were on our own just trying to figure it out. It was an enormously talented group of people who just didn't have a whole lot of experience. The creative energy flow was way more intense than Los Angeles, light years ahead in the depth of music played and in sheer insanity. At 'BCN it was like a playground; we were like little kids. We loved it, adored it, and had a passion for it.

Steve Segal became "the Seagull" on the air. "First I thought that Steve

Segal sounded way too Jewish and I wasn't comfortable being myself on the air yet. So I guess [the name] had something to do with me coming in from California." Don Law, busy running the Tea Party, became Segal's first Boston roommate: "Other than Ray Riepen, there is nobody more responsible for what happened at WBCN than Steve. They all followed his lead, everybody. He moved in with me on Beacon Hill and we were the odd couple. He was a really sweet guy, but he had a very tough time keeping it together. He was as disorganized as an adolescent, [with a] kind of arrested development, but he was brilliant, just brilliant."

"There's no question as I look back, but I was quite insane," Segal added; "I was clearly bi-polar. We knew I had something, but we weren't sure what it was. Manic depression sounded kind of right; it was a Jimi Hendrix song and it made some sense." Despite Segal's genius as a DJ, his personal life of constant chaos indicated that he might not be the best choice to lead the air staff, so Sam Kopper assumed that role. "When Steven arrived, he was the P.D. [program director], and the 'John Lennon' of our station. I basically exercised and made real his visions."

The term "program director" was not a real part of the WBCN vernacular yet, but someone had to sit in the middle and make adjustments, even if those decisions were pondered in billows of smoke around a hookah pipe or during the meandering circles of a passed joint. Indeed, in the June 1970 issue of *Boston* magazine, journalist (and future WBCN employee) David L. Bieber wrote an in-depth six-page article about the radio station, not mentioning the words "program director" once. He did, however, explain 'BCN's method for loosely organizing the anarchy of musical choice available to each jock: "Steve Segal refined and formulated the current station concept of programming as a train of thought via music. In this approach, two to four related records are connected by a sometimes fragile relationship which can be musical, thematic or consist of several different cuts which evoke a uniform feeling. Each of the station's announcers follow the formula." Joe Rogers commented on that concept in the article: "The music is all flow of moods, and, properly displayed, sparks can flash between two cuts." Creating montages of related sound bites, songs, and comedy pieces completely on the fly, thereby providing golden links in some kind of intellectual or even spiritual thread, made the station unique among its counterparts as the jocks strived for something greater than merely putting two songs back to back.

As Peter Wolf and Jim Parry stomped their feet in the Tea Party to stay warm, the last of winter finally transformed into a magical spring. Boston's underground radio experiment dropped its training wheels and began wobbling down the road in all of its groovy, hit or miss glory. The station's first promotional placard appeared: a pen-and-ink sketch of a lovely psychedelic lady peering out from her tresses to proclaim, "Ugly Radio Is Dead!" Record label representatives, swiftly realizing the opportunity to promote their hip underground albums through WBCN airplay, began dropping stacks of promotional copies off at the station. Rapidly growing record libraries were established at both Newbury Street and the back room at the Tea Party, sparing the jocks that long walk with heavy boxes of vinyl from their home collections (except for Wolf, of course, who still relied on his prized assortment of R & B singles). Parry remembered, "The record library [at the Tea Party] was four shelves in this closet in the back room. One night, I think it was when Ten Years After was there, Ray Riepen [walked in] and expansively told [guitarist] Alvin Lee, 'Hey, you want some records? Take anything you want.' So, he took the whole first shelf, which was everything A through Beatles, and like, the Doors! So, for the next month or so, we couldn't play any band whose name began with anything before the letter E!"

Back at 171 Newbury, "We had two Sparta boards tied together, five-channel each, with rotary pots, the cheapest you could get for a radio station," Sam Kopper recalled.

> But we had a [great] Neumann microphone and these two big old transcription turntables. The platters were, like, 17 inches across! They were really slow-starting, you couldn't just hit the start switch, so you had to slip-cue all the records [start the turntable, hold the disc steady on its felt pad while the turntable rotated below, and then let it fly when it was time to play the song]. Directly behind the disc-jockey was the outside wall and there was an air conditioner in there. In the summer of '68, our engineer, Sassy John "Ten Thumbs," who wore a long green raincoat like a flasher and never took it off, rigged up a relay so that whenever the microphone was on, the air conditioner would shut off. Therefore, every time the listeners would hear a [jock] start talking, in the background they'd also hear this "EEEEEEEEEUUUUUUUUU . . ." sound of the air conditioner [winding down]!

"Because the air conditioner was completely incapable of doing anything to keep the studio livable," Parry laughed, "we'd climb out onto the roof of the next building from a window in the production studio." Recalling the scene for *Record World* magazine in 1978, Parry added, "It was like looking over the rooftops of Paris. We'd get long cables on the microphones, run them out the window and broadcast from the roof as the sun was coming up."

Soon, a quarter-million students made their annual fall migration to Boston. After three months at home, many were delighted to hear that the hip, underground radio experiment they'd loved during spring semester had not gone away and, in fact, had thrived. WBCN's attic studio, dubbed the "Penthouse" by Wolf, no longer stifled its inhabitants, and the underwhelming air conditioner was shut down for the season. As the cold returned, the adjoining roof where the jocks had fled for relief from the heat now doubled as a wintry battlefield. "There were some infamous snowball fights out on that roof," Perry recollected with a laugh. "We had a big fight one night while Charles was on the air," referring to a brand-new jock named Charles Laquidara. "He made two or three snowballs and brought them into the studio [for defense], then put them on the board. Of course, they melted! The damn station went off the air!"

Laquidara rebutted this tale: "The real story is that a listener named Deirdre, who called herself 'Green,' came in from the street and handed me that infamous snowball with a green ribbon wrapped around it." But, the result was the same: the snowball was forgotten and the sound of static soon replaced that of music on 104.1. Laquidara, who began working at the station in December '68 and would become WBCN's most enduring, and endearing, personality for its entire history, was one of two significant hires. The other new voice was J.J. Jackson, who would achieve industry notoriety as the earliest radio proponent of an unknown new band named Led Zeppelin, as well as gaining nationwide popularity as one of MTV's first video jocks.

By day, Jackson held down a computer technology job to pay the bills, and then hit the clubs at night to explore his real passion: music. One evening he discovered the Hallucinations playing a show and, after striking up a conversation with the band's lead singer, found that Wolf doubled as a disc jockey on the all-new WBCN. A novice jock himself, on Tufts University's AM station WTUF (the facilities of which would soon be closed down by the FCC and occupied by WMFO-FM in January 1970), Jackson tuned in

Sam Kopper on the air at 171 Newbury Street, WBCN's first location.
Courtesy of the Sam Kopper Archives.

104.1 and was blown away. "I really flipped out over it," Jackson told *Record World* in June 1978. "I went up to visit Peter while he was on the air, and I just fell in love with the station and everyone I met. There was a lot of love and warmth there."

"I knew J.J. Jackson to be a 300-pound R & B singer," Wolf recalled. "So he called me from an old club called the Sugar Shack. I said, 'Well, come on by.' There was no one in the studio but me, so I ran downstairs, opened the door and there was this thin cat there, wearing 'the hat' and glasses: the whole 'Super-fly' outfit, pimped out. I realized, 'This is not the R & B singer.' He says, 'I love your show; can I come up?' He seemed genuinely into the station, so I said, 'Sure!'" Jackson's timing, like all of those in 'BCN's early lineup, was simply perfect, plus he was a black man. Jackson arrived at a time when diversity and "love your brother" were real aims. "Early on we wanted to hire a black guy," Sam Kopper explained, "plus, J.J. Jackson was one of the great souls and hearts going; he was just a sweet human being." The new guy was slotted into the WBCN's midday shift.

Steve Segal worked closely with the station's new hire: "I was the one that kind of mentored J.J.; I was the one who taught him to say 'ask' instead of 'acts.' J.J. was obviously a historical figure; he figured out how to be a commercial success and eventually became the first Afro-American VJ."

"The racial irony of the thing," Kopper revealed, "was that out of all of us, he probably played the least Motown and soul, and the most Led Zeppelin and Yardbirds." The evidence agrees: J.J. Jackson was acknowledged by guitarist Jimmy Page as being the first disc jockey in the *world* to play Zeppelin's second and third albums on the air. Laquidara recalled that the station also played a "white-label record" (advance promotional copy) of Led Zeppelin's first album when no one else had even heard of the group. J.J.'s own memorable moments included emcee spots for Led Zeppelin at the Boston Tea Party, Carousel Ballroom in Framingham, and Boston Garden. Pete Townshend joined J.J. Jackson in the 'BCN studio to personally debut The Who's *Tommy* on the air; the DJ hosted The Who at Boston University, and he brought Jimi Hendrix onstage at the Garden.

Any radio sales manager will tell you that the morning and afternoon drive shifts provide most of a station's bread and butter. In 1968, though, evenings and overnights were where the action was at WBCN. Wolf might have kept vampire hours, but his shift was key in achieving Riepen's advertising goal. But, as the year grew long in the tooth, "Woofuh-Goofuh" began having second thoughts. The Hallucinations had packed it in, and the singer fell in with a new group of musicians, the J. Geils Blues Band. As that group expanded its lineup and moved from acoustic to electric, the members dropped the "Blues" from their name and began attracting a following around town. Soon his growing commitment to the new project edged into Wolf's all-night radio turf. Sam Kopper related, "It was early December that Peter said he couldn't do a show five days a week anymore because [the J. Geils Band] was gigging all the time. He said he would do some kind of shift when he could."

Up to this point, the occasional missed shifts were covered by Wolf's crew, as Master Blaster revealed: "It was always his show, I was just hanging out. [But] sometimes when he'd go to a rehearsal or something, I'd have to do the show myself. I didn't really know what I was doing; it was flying by the seat of your pants, but it was cool." Steve Segal remembered, "When the band finally started to make it, Peter would miss his show, like, three or four times a week. He'd put the Master Blaster, the black jumpsuit guy, on instead. 'Hey' [imitating the sidekick], 'it's Master Blaster, gonna give it to you faster. All you young white girls, you ought to come on down!'" Segal chuckled at the memory: "I heard this [on the air] coming back from a jazz show in Newport and I was, like, 'No, no, no!' So, that's when we had to give Peter the option: you can either do overnights or you can become a rock star, but it's got to be your choice."

At this moment, in another example of perfect timing, Milford, Massachusetts, native Charles Laquidara arrived back in Boston after a long voyage of self-discovery that had taken him to the Rhode Island School of Design and out to Southern California, where he studied performing arts at the historic Pasadena Playhouse. After an intense period of taking acting classes, reading for auditions, acting in shows, and even writing and directing his own play, he graduated with a bachelor of theater arts degree in 1963. However, like most of the male members of his class, he now worried about being drafted. While that threat loomed, Laquidara's primary focus became a search for employment. In 1967, he auditioned for director David Fleischer in the lead role for *The Boston Strangler* but, as an unfamiliar face, lost that part to Tony Curtis (an event Laquidara would rue publicly on the air for his entire radio career). However, the frustrated actor did manage to find some work, a fortuitous path as it turned out, from an unexpected quarter: the basement of the Pasadena Presbyterian Church.

New recruits: J.J. Jackson and Charles Laquidara on Cambridge Common, 1969.
Photo by Sam Kopper.

Laquidara found out about the opportunity from his buddy Dave Pierce, another starving graduate who wanted more than anything to be a rock and roll disc jockey. He had settled for spinning classical records on the local radio station KPPC-FM, owned by the Presbyterian church, which programmed the music when it wasn't broadcasting Sunday services or midweek devotionals. Pierce, who knew very little about the genre, nevertheless took the gig and would listen to his favorite R & B records off the air while he spun the long symphonic pieces on the radio. He invited Laquidara, who knew slightly more about classical music, down to the station to help him out. "Dave Pierce taught me how to do radio, how to run the board, and he got me a job," Laquidara remembered. "I couldn't pronounce the names of the composers, so I had to be tutored." While the two gradually got a handle on how to suavely pronounce Shostakovich and Prokofiev like the pros, they managed to survive at KPPC while their station went through some rocky management changes, passing from a classical format to jazz, and then finally morphing into the second underground rock station in the country in 1967.

A year later, now a rock radio veteran with experience under his belt, Charles Laquidara had to head east when his mother died. Back in Milford, he tuned in WBCN, heard Jim Parry on the air, and called him up. When Parry noted Laquidara's local roots and West Coast experience, he encouraged him to visit the station, where the two met and Parry introduced him to fellow KPPC-FM alum Steve Segal. Now, as the 'BCN's overnight star was about to leave, Laquidara found himself in the perfect spot. Sam Kopper recounted: "I remember there was this [final] falling out between Peter Wolf and Ray. So, it went [quickly] from 'maybe we should bring this Laquidara dude in' to 'Laquidara is doing nights.'"

"The timing was amazing," Steve Segal mentioned. "Charles just walked in. I said, 'Sam and I just thought we'd throw you on the air and see what happens.' At the time, he was an actor playing an underground disc jockey, but over the years he became a knockout performer."

"I was a mediocre everything: a writer, cartoonist, actor, disc jockey," Laquidara explained. "However, in 1968, with the advent of underground radio, there was a place for a guy who was simply real. He didn't have to have a deep voice; he didn't have to talk fast or have a golden throat. And, he could fuck up, which I did . . . a lot! Sam Kopper went to Ray Riepen and he hired me. To this day I'm indebted to Ray Riepen." Then he added with a snicker, "Even though he tried to fire me a couple times."

I READ THE NEWS TODAY

By the beginning of 1969, the *Wheels of Fire* album by Cream, with its rambling, fifteen-minute rock jams and abstruse, poetic lyrics, had gone to number 1 in America; the "all-you-need-is-love" Beatles were squabbling, and Janis Joplin brought San Francisco acid blues to the top of the album chart. Martin Luther King Jr.'s path of nonviolent civil rights protest ended in gunfire and death; three Americans had just circled the moon in Apollo 8; the Vietnamese peace talks commenced in Paris; the Democratic National Convention had been ravaged by a torrent of street violence; and likely presidential nominee Bobby Kennedy had been assassinated in Los Angeles. "The jocks really knew their stuff," Ron Della Chiesa observed. "They were on the edge of what was new, what was happening, and what the youth market was going to be. Think of the timing of all that: Vietnam was going on, the protests, civil rights, the spin-offs from the assassinations, the country in upheaval. Then, there was the music; it couldn't have been more timely . . . the stars lined up." Tommy Hadges remembered, "There was an amazing array of music that was coming out. The war was going on; there was a cultural revolution, a drug revolution, a political revolution. But [it was] also about being in Boston . . . in a city where it is renewed and refreshed every year with all the new students that come in. With the university environment having a lot to do with the social revolution going on, WBCN fit right into that."

Joe Rogers offered his view of WBCN's music mission: "I always felt that I was there to bring this music to the people. The reflex was, you've got this huge record library and 'look at the things I've found!' The emphasis was on doing sets [of music] and segues; we thought that's what our craft was. You tried to sneak one song into another in a clever way, whatever that would mean."

"It was kind of like college radio, you played the music you liked and you talked about it," Jim Parry described. "Each show was quite a bit different. I would put a lot of folkie things in and Charles, at one point, was fired because he was too 'rocky.' That lasted a couple of days and then he was back," he laughed. "We all pretty much winged it." Tom Gamache (known as "Uncle T"), who eventually got on the air at 'BCN in March 1969, thrived on the spontaneity: "I decide what I'm going to play about two minutes before I put it on the turntable," he told the *Boston Globe* that same month. A lot of Gamache's choices were "the most bizarre," according to Laquidara. "He blew our minds with Frank Zappa and the Mothers; he turned us all onto 'Witchi Tai To'; he was the guy that played Captain Beefheart, John Coltrane, and Roland Kirk. If every station has that guy that pulls the most brilliant songs out of his ass, he was that guy." J.J. Jackson told *Record World* magazine in 1978, "The jock was allowed to show his personality on the air, and lay out the show the way they wanted. You could play everything from Stockhausen to Alvin Lee; the only record you knew you were going to play was the first one."

"It was a very mellow presentation," Sam Kopper described. "A lot of times we were stoned on the air and we came across very gently, very conversationally; that's one hallmark of that time. The other thing I really give due credit to Steven [Segal] and then Charles, was the madness that lasted at 'BCN into the early nineties. That was laid down at the very beginning." Segal found the on-air lunacy to be quite normal because "there was so much bizarre stuff happening in real life! For me, I'd just kind of start the engine and see what came up. It was almost always spontaneous; I never knew what I was going to say on the next break." As good as he was, "the Seagull" was still inspired by the DJ who became one of his best friends, Joe Rogers (as Mississippi Harold Wilson on the air). "Joe did things with nuance, things that came in from completely out in left field. He was a soft-spoken guy who didn't have a mean bone in his body." He laughed as he recalled an example:

> Mississippi had Ian Anderson on; [Jethro Tull] was at the Tea Party and he had to come to 'BCN for an interview. Ian was a totally unlikable person, an unbelievably arrogant human being, and he was really nasty to Joe. I heard them going on for a few minutes and I remember coming in after Ian had finished being sarcastic and smarmy, and over Joe's shoulder I said something into the mike like, "Are you this nasty to all of the disc jockeys who play your music so people will know what you're doing? Could you ever come in and just answer a question straight without any attitude or acting superior?" For a moment I think he had the starch taken out of him! To this day I know Joe would say that's the way a lot of musicians were, and it's true, but it didn't make it a good thing to do. That was my favorite: basically calling Ian Anderson an asshole.

He added with a snicker, "Did I get that one right or what!"

Early in 1969, the fledgling WBCN air staff lost two of its own. A founding rock and roll father in "The American Revolution," Tommy Hadges, became an absentee jock, only showing occasionally for fill-in shifts, now that he had decided to concentrate on his studies at Tufts Dental School. "Yeah, going back to Dental School, that was a great idea," Steve Segal kidded glibly. "I said to him, 'I can't believe you're doing this!'" But shortly after Hadges's exit, the West Coast guru himself defected, heading back to Los Angeles to be a pioneer on the new underground outlet KMET-FM. That promising experience would prove to be so unsatisfying and "corporate," according to the jock, that he left after only a few months and ended up back on KPPC in Pasadena. With Segal's sudden absence, Charles, who had moved in to replace Peter Wolf, now assumed an earlier time slot, while Uncle T and Jim Parry handled the late shifts. As of April 1969, the weekday lineup had shaken out to 7:00 to 10:00 a.m.: Sam Kopper; 10:00 a.m. to 2:00 p.m.: J. J. Jackson; 2:00 to 6:00 p.m.: Mississippi Harold Wilson;

6:00 to 10:00 p.m.: Charles Laquidara; 10:00 p.m. to 2:00 a.m.: Uncle T; and 2:00 to 5:00 a.m. (until the smattering of taped religious broadcasting began) was handled by cleanup man Jim Parry.

Al Perry, who still managed to avoid a good night's sleep by jocking on the weekends and doing overnight fill-ins, was now out on the streets selling ad spots most of the time. Perry, who had become sales manager, recalled that "in the beginning there was me and Jack [Kearney]—couldn't have done it without him. He was one of the financial cinderblocks of that place, as was Kenny [Greenblatt], who was a real character." Tim Montgomery, who joined the station two years later and eventually became a sales manager, remembered Greenblatt fondly.

> Kenny's job was to interface with the record companies, so his job, to tell you the truth, was almost as important as the program director. He lived in the heart of Harvard Square next door to Peter Wolf, and the only thing he did was hang out with [record company] promo people. He was out and about every night, seven nights a week . . . which is, unfortunately, why he's not with us anymore! He wore moccasins, had long hair, was fond of saying that everything was "far out." He would twirl his handlebar moustache and say, "Timmy, Timmy, you've got to hear this! It's heavy, it's *far out!*"

The members of the fledgling WBCN sales department, as sleepless or as consciousness altered as they might have been, were kicking butt. Ernie Santosuosso noted that fact in his 1969 *Boston Globe* article: "The station now has almost $25,000 in ad billings a month, the second-highest for an FM station in this area. WJIB-FM, an 'easy-listening' station, is first." Al Perry and his crew achieved this despite the rule set by Ray Riepen that 'BCN's commercial load was not to exceed eight ads an hour. Perry noted, "No one believed we could make it; we were the underdogs, not just from a radio standpoint, but a musical one too. All of FM radio only had a 10 [percent] share [of the Boston audience]; WRKO-AM was the king." That AM Top 40 giant alone often pulled in twice the ratings of the entire FM band. But even if WRKO possessed massive numbers, WBCN still had the guns. Al Perry elaborated, "Jack Kearney, J. J. Jeffries—who was a jock at 'RKO, and I, we'd all go out drinking, and J.J. would say, 'I was getting laid last night [while] listening to 'BCN!' No one cool listened to 'RKO.'"

Apparently, though, there were a lot of cool people listening in to 'BCN,

Kenny Greenblatt with Jim Parry on Cambridge Common. Greenblatt worked the record labels for 'BCN and found an enthusiastic source of income. Photo by Sam Kopper.

many of them college students whom the Arbitron ratings service did not survey because of their transient nature. Charles Laquidara recalled, "You could walk from one end of Boston to the other, from Stuart Street all the way to Cambridge, and you wouldn't need a radio because every dormitory, every apartment, all the stores, would have 'BCN blasting out the windows." If the station featured a brand-new group or a song that hadn't been released as a single yet, chances were that people were hearing it first on WBCN. Sam Kopper told *Boston* magazine in 1970 that over a year earlier, he had witnessed clear evidence of the station's influence with two releases. "The distributor brought us a tape of Traffic's second album four weeks before the record came out. We were the only station playing Traffic, and when the album was released, it completely sold out locally in a few days. After similar prerelease play of Led Zeppelin's first album, the local stores received their shipments in the morning and all copies were gone by afternoon."

Business was going so well at the end of 1968 that Ray Riepen managed to convince Mitch Hastings to give the DJs a raise, as Sam Kopper revealed: "We all started at $85 a week, and by early '69 or so, all the jocks were making $125 a week!" Riepen also successfully pointed out the need to move into bigger digs. Jim Parry elaborated, "Mitch had this dream of having a real radio station on Newbury Street, with that prestigious address, but it wasn't really viable as a station; we ran out of space."

"We moved to 312 Stuart Street [which] was just above Flash's Snack and Soda and down from Trinity Liquors," Sam Kopper added. "Charles went to work just as we moved, right at the end of 1968." The new location, nestled behind the bustling Greyhound Bus terminal, would be WBCN's home until June 1973. Flimsy and cheap quarter-inch paneling notwithstanding, the layout was much more conducive to running a radio business, with accessible areas for reception, sales, and management; studios in the back; and the record library across the hall. Perhaps best of all, the DJs had escaped their claustrophobic Newbury Street attic; they wouldn't freeze during the winter months or roast in their own sweat next to a loud and impotent air conditioner during the summer.

As WBCN settled in on Stuart Street, some fresh voices appeared on the air. John Brodey, who would eventually become a radio star in his own right and a future music director, had listened to 'BCN with great interest during his frequent visits home from the University of Wisconsin to see his parents in Boston. "I knew a girl who had gone out with Steven [Segal] for a while and I said, 'If you can just give me an introduction, I'll take it from there.' Soon I was interning for Steven, weaseling my way in. I had to put away his records, get him stuff, and drive him home because he had a car but didn't drive. We started talking, and soon it was: 'Oh, you know something about music, man.' It was a nice relationship and . . . I lasted!"

Sam Kopper also hired Debbie Ullman to be WBCN's first female jock after she had worked with the close-knit crew in the sales department for a time and also volunteered to helped transport the station's voluminous record library from Newbury to Stuart Street. "I liked her voice, intelligence, and spirit," Kopper mentioned. He was also impressed when she took the initiative to study for and pass the FCC exam for a radio license, not necessarily an easy thing to do, and unusual, at the time, for a female. "I was the only woman out of about forty taking the exam; afterwards the guy looked over my test and was startled." So now she had the endorsement of Uncle Sam, but did she have the chutzpah to do a show? Kopper decided to find out. As Ullman worked in the sales area one afternoon, the jocks arrived for an air staff meeting. "That usually meant that someone would play long tracks on the radio, like a Grateful Dead side, while they met," Ullman remembered. "On this occasion someone came [to me] and said, 'We're going to have a meeting in ten minutes; if you can learn the [air studio] board, you're on." She wasn't thrown by the challenge. "I'd already been watching [the other jocks use] the board . . . so that wasn't hard. I went on while they had their staff meeting; apparently, I didn't fumble things too badly!"

Sam Kopper also hired Andy Beaubien, a recent University of Rhode Island grad who had begun working in local radio when he was only sixteen and did shifts at some small Rhode Island stations throughout his schooling. "As I was wrapping up my college years, I was saying to myself, 'I'm getting kind of tired of this radio thing.'" Beaubien laughed because, as it turned out, being a DJ and radio programmer would become his lifelong career. "There were only two stations I was really interested in working [at]. One was WNEW-FM in New York and the other was 'BCN. I asked Sam if he had any openings, and he said, 'If you want to start working part time, you can.' The weekend I started was the weekend of [the] Woodstock [festival], August 1969, so that's why I didn't go [to the concert]. I worked there for about a month or two; then Steven left and created an opening." Beaubien recalled that Ray Riepen wanted to meet him when Kopper recommended him for the full-time position.

> I was very intimidated because I had heard all the stories about Ray Riepen. I walked in there and he said, "Well, tell me about yourself." So I gave him my story, and then he said, in his Kansas City accent, "You know, there's only one thing about you that kind of worries me." I said, "What's that?" "You've got previous radio experience. I prefer people who aren't tainted by commercial radio." I had to explain to him the reason I wanted to work at 'BCN was [that], simply, I didn't want to have anything to do with typical radio and that I wanted to get into the music. Obviously, I convinced him because he gave his blessing and I ended up working there.

Beaubien immediately fell in love with his job.

> It was the most fun radio experience I ever had, even more fun than college radio. The station was truly free-form, no restrictions at all. The only thing we tried not to play was something we'd done in the previous couple of shows; the idea of repeating something was bad. We would have meetings

once a week and sit around and talk about various issues, including the music. We tried, in a kind of informal, unstructured way, to keep a certain level of consistency in the programming. In other words, if somebody was going too far off the deep end . . . let's say, playing too much John Coltrane, someone might say: "Hey, it's cool to play Coltrane, but you don't want to do three jazz artists in a row." All of us had our predilections. Charles liked to play a classical piece every so often and we'd have discussions about that. We'd say, "Charles, when you played that Brahms symphony, it really sounded so . . . unlike us!"

Chuckling to himself, he continued, "Jim Parry was really into folk music and acoustic blues—that was his predilection, and sometimes he'd go off the deep end there. I was really into guitarists in those days, so I was liable to go off in an Eric Clapton and Django Reinhardt set."

"If you can imagine eight or nine people sitting down in a music meeting for an hour and trying to agree on anything—it was impossible!" Tommy Hadges laughed. "But . . . it was also wonderful . . . to have that freedom. It just isn't anything that exists in broadcast radio today."

Music epitomized the main message of the early WBCN, but a voice of conscience always ran close alongside. The DJs didn't attempt to separate music from their politics, which they wore on their sleeve. Ernie Santo-suosso, in a March 1969 *Boston Globe* article entitled "The Beautiful Radio People," wrote, "In an industry saturated by news broadcasts, WBCN is predictably unique. The announcer often allows the record of the moment to serve as his vehicle for a solitary news item." Sam Kopper told Santosuosso how he handled the moments leading up to Senator Robert Kennedy's death: "I read the hospital bulletin during the playing of 'Come On, People, Let's Get Together' [the Youngbloods' "Get Together"] and just before the beginning of the vocal. It fitted beautifully." Steve Segal described in the article about how he reacted to the scenes of street fighting at the Democratic Convention: "I couldn't refrain from talking about the Chicago violence. When I saw something as blatantly ridiculous . . . and friends being clubbed, I spoke about five words and put on [Everett] Dirksen reading the Declaration of Independence." Everett Dirksen was a Republican senator from Illinois, the Senate Minority leader, and a staunch supporter of the Vietnam War. Years later, Segal clarified, "He had [put out] an album of folksy, yet arrogant, readings of the Declaration of Independence and other

Sam Kopper at WBCN's second location on Stuart Street. "Above Flash's Snack and Soda and down from Trinity Liquors." Photo by Don Sanford.

documents of freedom over muted patriotic music. That this puss bag had put these amazing words on vinyl for profit at such a time of government suppression and violence toward its own people seemed so bitterly ironic and innately satiric on so many levels, I figured not a whole lot needed to be said."

For the first year and a half after WBCN's underground radio transformation, there was no formal news department at the station. However, discussions about headlines and politics on the air were part of each DJ's show. "It was important to let people know where they could get advice on what to do about the draft and the war," Joe Rogers remembered. "The station insisted that [listeners] become informed on the subject; if you

were pressed up against choices, that you knew what your choices were."
Ray Riepen did not restrict his staff from expressing themselves politically:
"That was part of the deal . . . that's what we were. We didn't take any radical
rant and rave stuff [on the air] because I wanted them to play music, but
we *were* against the war in Vietnam and nobody else [in the media] was
yet. A lot of people were upset that we weren't toeing the line as far as the
war, but that's what we were all about." Debbie Ullman acknowledged,
"The 'programming' of the music was definitely about fun . . . and hedo-
nism and all of that. But there was also a very strong visionary sense. We
actually saw the civil rights movement come to some degree of fruition,
we were in the process of bringing the Vietnam War down . . . there was a
tremendous sense that we could change the world. We thought we were
creating a model of working together and a more peaceful world. It wasn't
a spectator sport."

Al Perry added soberly,

We had a sense that something was wrong in America, with the war and
Nixon. It had to change and we were a part of that. I think it was when we
went into Cambodia, everything had just erupted—Harvard Square was a
mess, kids were demonstrating. I remember this guy, I believe he worked
for Hancock on Stuart Street, came up the elevator. I was there early in the
morning and I said, "Can I help you?"

He said, "I was listening to [BCN] last night and it's outrageous what you
do and say on the radio. I'm going to call the FCC."

I said, "Well, what are radio licenses for?"

"What?"

"What are FCC licenses for? They're to serve the community. If you had been
here at two o'clock this morning and seen the phones ringing off the hook
and a bunch of volunteers answering them because [the United States] just
invaded Cambodia . . . who are you kidding?" He got pissed off and left. But
that's it, isn't it? We served the public. Maybe we didn't serve all the public,
but we served *our* public.

In October 1969 WBCN took its political pursuits one step further by
creating its first news department, but it all happened in a rather round-
about way. Sam Kopper tapped Brooklyn native and Brandeis student
Norm Winer to be WBCN's latest part-timer after the twenty-one-year-old

impressed him with his knowledge of music. Winer remembered, "Andy Beaubien joined the station at about the same time, and he and I were both going for Steven [Segal's] gig. Andy got [it], so they tried to find me something to do around the station. Since I read the *New York Times* every day, they hired me to be the first news director."

There had been no newswire service at WBCN since United Press International removed its teletype machine years earlier. Since then, any headlines read by the jocks came from a newspaper, so breaking stories on the air was out of the question. In response to the audience's hunger for information, especially about its own government and also the Vietnam War, as well as to satisfy FCC requirements to broadcast a certain amount of news and public affairs programming, Ray Riepen arranged a contract with Reuters. The venerable, century-old newswire service from London installed one of its teletype machines in the Stuart Street headquarters, and soon reams of paper were cascading out of the chugging contraption, providing the jocks with what they referred to as "rip and read" news: available for their shows even earlier than the newspapers could report it. There was some significance to having a foreign news service supply WBCN's information as opposed to the U.S.-based companies, as Joe Rogers pointed out: "Reuters from England had some perspective on world affairs, it was not just rote [information]." Danny Schechter, who would soon arrive and make a memorable mark as 'BCN's principal news reporter, pointed out in his book *The More You Watch, The Less You Know* that there was also a more pragmatic reason for installing the English teletype: Reuters offered its services at a cheaper rate than Associated Press or United Press International, so Riepen saved the ever-parsimonious T. Mitchell Hastings a bit of cash.

WBCN now had an official news department, even if it only possessed one employee, but Norm Winer tore into his responsibilities with enthusiasm. Pursuing interviews and sound bites to spice up his segments on the air, he'd regularly head out to the rally, march, or sit-in of the day. "When all the demonstrations and protests were happening, I would bring an old tape recorder with me. I'd get arrested, then reveal my press credentials at the very last minute so that I could get back to the station and broadcast the news!" This pattern—of finding creative ways to present what, in most media outlets, was a bland concoction of terse stories—remained a mission to be continued for all the years that WBCN presented news as part of its programming commitment. As Danny Schechter summarized in his book: "If the music was going to be different, why not the news?" But Winer, who established the station's attitude of going out to embrace the stories, retained the news director position for only seven months. In a surprise development, Steve Segal enticed his good friend Joe Rogers to join him at KPPC-FM. After a flurry of tearful send-offs from the glum air staff, Mississippi Harold Wilson signed off at WBCN in June 1970, emerging days later in California with a sunnier alias: Mississippi Brian Wilson (but that's another story). In the wake of the abrupt change, Norm Winer got his full-time DJ shift, moving to overnights, while a new figure was recruited to handle the young news department. His tenure, though, would be even briefer than Winer's: a mere four days, in fact. But it's one of the most legendary in the WBCN scrapbook and an emblem of the turbulent times that existed at that point in America.

Ray Riepen supervised some interviews for the news director position, and one of those applicants was Bo Burlingham, a politically aware activist who, like many of his young generation, was catalyzed by the government's activities in Southeast Asia. He had been a member of the SDS (Students for a Democratic Society), which organized protests on campuses across America. Burlingham had also been part of the Weather Underground, a radical offshoot of SDS dedicated to establishing a revolutionary party to take over the U.S. government, even with violent means, if necessary, to return power to its people. In 1969, the Weathermen, as they were also known, sent a small group of followers to Cuba to meet with communist officials from that country, as well as from North Vietnam, to discuss America's political opposition to the war. While the Weathermen considered themselves to be patriots opposing the government's illegal and warlike actions, the leaders in Washington considered it high treason. "I had been a member of the Weather Underground in the early days, in the Central Committee," Burlingham admitted. "I went to Cuba with the SDS delegation, but I left them pretty soon afterward. My wife and I needed to find a place to live and we settled on Boston." He heard about the WBCN job opening from a mutual friend of Charles Laquidara's and came in for an interview. "Ray Riepen was quite impressed that I had graduated from Princeton—even though I had barely managed that." While the sheepskin from New Jersey scored points in his favor, Burlingham didn't consider his political activities relevant enough to bring up, so as far as Riepen was concerned, he had found an outstanding candidate and promptly gave him the job.

Bo Burlingham was now a happy man. He had a position that suited his interests, at a radio station whose employees were clearly not conservative or even mainstream in their political leanings. "I started on a Monday as news director. We would take news stories and insert sound and music segments. I was so slow at it at first that we had a slogan for it: 'Yesterday's stories—tomorrow!' Tuesday we had improved and by Wednesday we were just hitting our stride. Then on Thursday, the teletype went off. I went over to check the machine and noticed the first word said 'BULLETIN.' So, this had to be important, but you had to wait while it typed out each line." As Burlingham read the words clacking out on the machine below, his body began to chill. "The title was something like, 'FEDERAL GRAND JURY IN DETROIT INDICTS 13 MEMBERS OF THE RADICAL GROUP, THE WEATHER-MEN, ON CHARGES OF CONSPIRACY.' All the first names were people from the leadership, people I knew, some of whom were underground by this point. Then the teletype gets to the last name and I saw that it was . . . me! I just reached down, ripped off the story, folded it and put it in my pocket." Numbly, Burlingham walked into the next room, bumped into Laquidara, and handed him the incriminating evidence, but he didn't get a whole lot of sympathy at first: "He reads the story and the first thing he's worried about is where they're going to get a new news director! I don't know if it was him or someone else who said, 'You're going to need a lawyer.'"

Bo Burlingham confessed to his membership in the Weather Under-ground but asserted that he did not serve in any critical, policy-making role. "I was in and out very quickly. But, there had been a mole in the organiza-tion, a person whom I befriended and got to know. I guess I was indicted because they needed to enter in testimony from me at the trial." Danny Schechter revealed, "It was part of an FBI effort [called] 'COINTELPRO,' [an operation] to stop various protest movements in America by infiltrating them." Burlingham contacted his buddy, Michael Ansara, who had been the head of the Harvard chapter of SDS and the one who had originally tipped him off about the 'BCN job; then the two sped off to Cambridge on Ansara's motorcycle to break the news to Riepen. "We stopped in Kendall Square for something to eat and the story came on the television, naming all the names and showing pictures. I started to look down to try to hide my face; I could just picture my photo on the screen and someone across the room shouting, 'That's him!' But, [luckily] they didn't show a picture." After telling Riepen about the dilemma, the mercurial manager's advice was also, "Get a lawyer," which Burlingham did immediately. "[The lawyer] asked me whether I was going to turn myself in or go into hiding. I said I'd turn myself in, so he arranged for me to do that at the courthouse. The judge posted bail at $100,000 with lots of conditions: I had to surrender my passport, had to check in [regularly], advise them of any travel plans, and I couldn't drive through Rhode Island for some reason!"

While Bo Burlingham waited for a trial, which wouldn't begin for three more years, Riepen tried to figure out what to do with the alleged co-conspirator. The often-antagonistic comments about the government spoken on the air at WBCN didn't necessarily endear the station with local and federal authorities, plus Riepen believed that a couple of break-ins at his apartment were not just burglaries but the results of a government investigation into Boston's radical movement. "When [Burlingham] got indicted," the general manager divulged, "I said, 'I've got a federally licensed business here, and I've had enough heat from these guys.'"

"Ray decided that perhaps it would be best if I left 'BCN," Burlingham added.

"So, I gave him a job at my newspaper," Riepen said. Earlier that year, the entrepreneur had amassed some of his earnings from the Boston Tea Party and WBCN, joined with a financial partner, and invested in a tiny rag called the *Cambridge Phoenix*. Infusing the publication with cash and beefing up the editorial and reporting staff, their goal was to make a serious run at becoming Boston's weekly of choice. In a noble gesture, Riepen steered Burlingham toward this new opportunity. "He hooked me up with Harper Barnes, his editor, and I wrote stories for the *Cambridge Phoenix* under my wife Lisa's name."

When the trial of the Weather Underground leaders finally got to court in 1973, it was asserted that the government had obtained information about the defendants by wire taps and other potentially illegal surveillance. With the Supreme Court prohibiting electronic eavesdropping without first obtaining a court order, prosecutors realized they would be vulnerable in their efforts to win the case, since at least some of their critical information had been gathered by questionable means. Subsequently, the suit fell apart, and the charges were eventually dropped. Burlingham got his passport back and his name cleared, and then went on to flourish in a journalism career for many years to come. Despite his many successes, though, he'd never forget his four days working at WBCN in 1970.

In the sixties and early seventies, many policy makers in Washington considered the protest activities of its citizens as an outright attack on the country. With the Weathermen declaring an official state of war on the U.S. government in 1970, it was clear that the gloves had been completely taken off on both sides. Joe Rogers talked about the feeling of that time: "You know how you can be young and paranoid . . . but there was a sense that something politically important was going on and that you might very well be being watched. Who knew if the FBI would come storming through the door." Andy Beaubien added, "This may seem naive now, but there was a group of us at [WBCN] who really believed we were on the verge of a political revolution in the U.S. and if you listen to an album like *Volunteers* by the Jefferson Airplane, it expressed the mood of the time. The station went through a very political stage around 1970, '71, '72. We were pretty far out on the left and a lot of what we were doing on the air had to do with the politics of the time . . . this was Vietnam and a little later it was Watergate." In amazement, he recalled, "I remember emceeing a concert at BU; the artist was Buddy Miles. It was a white college audience and it was a benefit for the Black Panthers!"

Danny Schechter, the man chosen to replace Bo Burlingham, would be the asset that drove WBCN squarely into the political stream. By the time he arrived in Boston, he had already graduated from Cornell University and gotten his master's at the London School of Economics, but working in an environment that combined his pursuit of journalism and love of music seemed like the perfect place to be. At first, though, despite all of his schooling, and some journalism experience, Schechter began at the bottom as an intern working for his predecessor. "News at that time consisted of some headlines and one produced feature that Charles read and Bo wrote," he remembered. "But people said that Bo didn't have any radio experience and was going to have to produce an expanded newscast. Did I want to help out? So I did it first on a volunteer basis because I was interested in radio news, which I knew nothing about. It was the blind leading the blind, in a way!" He had barely started learning the ropes before Burlingham's indictment and sudden exit elevated the young intern into the principal news position. "There was, at the time, an advertising campaign that said, 'I got my job through the *New York Times*,' so I joked that, 'I got my job at 'BCN through the FBI!'"

It wasn't long before Schechter adopted the famous radio handle that

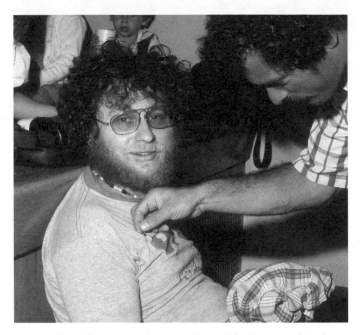

On a break from giving the news, Danny Schechter gives blood.
Photo by Dan Beach.

would remain with him throughout his career. "I was writing news [stories] for the DJs to read and I wasn't the best of typists, so I would correct the copy with a pen. [But] my penmanship was worse than my typing! At one point I handed Jim Parry some news which, typically, had my scrawl on it. He goes, 'What's this? I can't read this! *You* read this and *I'm* going to take a leak.'" Parry got behind the microphone and gave Schechter an impromptu windup for his very first radio moment: "He said something like, 'Now ladies and gentlemen, I'm going to introduce Danny Schechter—the news detector . . . the news inspector . . . the NEWS DISSECTOR!!' At which point I read the news, kind of terrified . . . then ended up staying there for ten years. So, thanks to Jim Parry, I adopted this News Dissector brand, if you will, and used that as a way to differentiate what I was doing from what others were doing in Boston."

The News Dissector lost his shakiness and grew in his craft.

We were playing to an audience in Boston, particularly three hundred thousand college students, at a time when the counterculture was in ascendancy and the antiwar movement was, in many ways, based in [the city] with Howard Zinn, Noam Chomsky, and other intellectuals. This was something that was popular; in 1972, the whole country was a landslide for Richard Nixon, except for Massachusetts. We did a radio show then called "Nixon: 49, America: 1," which was about that election. When most people think of politics they think of Democrats and Republicans and elected officials. That's not how we defined politics. We were really talking about politics in the community and on the streets . . . politics of protest and culture. 'BCN was, sort of, charging in another direction. We weren't lecturing the audience; it was [more] like engaging the audience.

Danny Schechter began to build up a department that could handle the amount of story investigation and production time needed to create two regular news broadcasts at 6:00 p.m. and 10:00 p.m., plus his frequent special reports. Bill Lichtenstein, who started as a volunteer in 1970, recalled that WBCN excelled at in-depth coverage but eschewed the "on the hour" punctuality that characterized all of the other television or radio news reporters in the area. "The six o'clock [news] would go on the air sometime between 6:00 and 7:30, and could run anywhere from twenty-five minutes to an hour. The ten o'clock news would go on anytime between 10:05 [and] 11:00 and run from fifteen to forty-five minutes, depending on what was going on." He added, with a laugh, "If Spiro Agnew was in town and there was a full-scale riot with people [being] arrested and we were getting phone calls from them, we'd be editing things together and Danny would be going into the studio with thirty-two carts [tape cartridges with recorded segments on them]. Charles would say, 'We've got a whole report coming up at 11:00, so stay tuned.' And *that* would be the ten o'clock news."

Like Schechter and most of the WBCN air staff before him, Lichtenstein was thrown in the deep end suddenly. He arrived at the station as a fourteen-year-old junior high student from Newton, which had a curriculum requiring its students to go into the community and find a volunteer job. "I called 'BCN and they had just started the Listener Line. The person I spoke to, Kate Curran, who[m] Charles brought in to set up this system to handle all the phone calls, swears that I said I was sixteen. So, I started answering the phone, in the days before the Internet and Google; [WBCN] prided themselves in the fact that you could find out anything! They stocked bookcases with reference books because people would call up with all kinds of questions . . . like 'What's the population of Sri Lanka?'"

Bill Lichtenstein, WBCN's youngest employee, covered a Black Panther story at age 14. Photo by Don Sanford.

One day, the fourteen-year-old—ahem, the *sixteen-year-old*—fell under the gaze of a stressed-out Schechter, as the volunteer worked one of his many Listener Line shifts.

> There was a big protest at the Boston Police station that used to be right up the street from Stuart Street. The police in Chicago had shot a Black Panther in his apartment and people were saying it was an assassination. Danny handed me a tape recorder. "Here, go up the street and cover this demonstration."
> "How do I do that?"
> "This is the microphone, push the red and white buttons and it'll record; then just ask people, 'Why are you here?'" So, I went up the street and people gave me all these really intense answers, [which] I brought back and he used for the news.

Thus began Bill Lichtenstein's long career as an award-winning investigative reporter and journalist, although shortly afterward he lost his cool, high school–mandated volunteer job when Schechter began paying him out of the news department budget.

Lichtenstein once covered a Boston demonstration and subsequent protest march through the city streets. He tagged along as the throng crossed the Charles River on the Massachusetts Avenue Bridge, heading into Cambridge. "It ended up at the Center for International Affairs over at Harvard. This was Kissinger's old office and it was tied to the war effort. They would meticulously call it the CFIA, but people would roll their eyes and say, 'No, the Center for International Affairs should be [called] the CIA!' [The crowd] ended up breaking in and trashing the place; they started ransacking files and tearing everything up." He laughed at the memory of finding a desk that hadn't yet been overturned or destroyed in the melee, which still had an intact telephone. "So, I called Danny up, live on the air from inside, to do a report while all this was going on." He was in the middle of his call to WBCN when suddenly, "Somebody yelled that we had to get out, the police were coming! I concluded my report with, 'Desks are being overturned, files are being ransacked and phones are being ripped out of the walls!' Then I pulled my plastic phone jack out of the wall too!"

Danny Schechter elevated the editorial power of the news department with two additions, Andy Kopkind and John Scagliotti. The former, a noted writer for left-wing magazines like the *New Republic* and *The Nation*, would later compose political essays in publications as prominent and respected as *Time* magazine. Before he got to WBCN, Kopkind worked in Washington at the Unicorn News Collective, which fed reports about issues concerning the war to counterculture radio stations across the country. His partner, John Scagliotti, revealed, "The collective, though, was going bad; it was time to leave D.C." Charles Laquidara, who knew Kopkind, invited him to come up to Boston. Scagliotti continued, "Andy went up, I followed, and while we were at 'BCN, Danny said, 'Why don't you do some stuff and maybe they'll hire you at some point.' It was all very flexible." So, despite Kopkind's semilegendary status, he and Scagliotti began on a volunteer basis, like all the others.

"Our very first ['BCN] piece was on the Victory Gardens in the Fenway," Scagliotti recalled. "No one knew what they were back in those days; [it was] just these people planting nice vegetables. This was the first piece of radio craziness that we did: at the beginning [of the report] we took *The Messiah*, the part where they sing, 'let us rejoice,' and we took all the 'let us' . . . 'let us' . . . 'let us' [parts] out and cut them all together to make 'lettuce . . . lettuce . . . lettuce.' Then we had [a recording of] a woman who just mentioned, 'We grow lettuce over here,' and then in comes the 'lettuce . . . lettuce . . . lettuce . . .' That was our early work; weird!" he added, chuckling. "In fact, we were discovering and developing a whole new sound that was beginning in the early seventies, because radio had never really mixed sound effects and live actualities in news and public affairs, like using a car crash sound when there was something [in the report] we didn't like." The chemistry clicked; soon Kopkind and his young protégé were both drawing checks from WBCN. Eventually, Scagliotti would even be made the news director, technically Schechter's superior. He downplayed this role: "I was the boss only because nobody else wanted to do it. When [the department] got big, somebody had to manage it and make sure everybody got paid!"

But, even though the news department allowed itself to smile, even laugh at times, by no means did its members forget that they were very much under the U.S. government's steady gaze, as Danny Schechter described: "There had been a vocal group of activists who put a bomb in the Suffolk County Courthouse. They issued a communiqué and then called 'BCN to say it was pasted up in a phone booth. We brought it to the station and I [went on the air and] reported it." The document ended up on Schechter's desk, crammed in a colossal pile as papers chronicling the next day's news events quickly plowed it under. Although a brilliant news man, Schechter wasn't, by any account, a neat one. "Then the FBI showed up and they wanted [the communiqué]. I didn't want to give it to them, but Al Perry [now WBCN's general manager] told me, 'You have to!' Everybody was living in fear of the FBI or the government yanking our [radio station] license. So I said, 'Okay, I'll give it to them.' But then I couldn't find it."

"You have to picture his office," Lichtenstein added with a snicker. "Danny would sit and clip newspapers and the clips would pile up, and the newspapers that were clipped would just pile up; the next day he would just start again. He had an enormous amount of material laying around."

"There was a half-eaten tuna fish sandwich or something on my desk; it was a mess!" Schechter apologized to the agents: "Honest, it's here someplace." After twenty minutes of watching him search, they became restless and disgusted, handing the news man a business card that he promptly "misplaced" too. The sensitive document never turned up again, but luckily for 'BCN, neither did the FBI men.

MOVIN' ON UP

"Media Freaks Act Out Battles of the Radicals" read the headline of the
Boston Globe story by Parker Donham in June 1970. He was describing the
wild, often naked, scenes at Goddard College in Plainfield, Vermont, as it
hosted the Alternative Media Conference First Gathering. The event was
a rallying cry for more than 1,500 hippie announcers, writers, producers,
and directors scattered across the country in the days before radio conferences became big business and MTV, and then the Internet, linked (and
homogenized) singular cultures across North America. The conference
represented a microcosm of the entire counterculture as different groups
with widely divergent views squared off in debate while extracurricular sex
and drug use flourished openly. A young Mark Parenteau, then a teenage
DJ in Worcester with the radio handle of Scotty Wainwright, signed up and
made his way north. "They were having *Creem* magazine, *Rolling Stone*,
other alternative newspapers, all the legendary FM jocks who had become

famous ... I guess it was a 'getting-together,' but it was [also] a huge party. Every city had a band representing them, and the band for Detroit was the MC5. Wow! I loved their energy. So that weekend I gravitated and hung out with all the Detroit people." That would lead to "Scotty Wainwright" meeting all the right names and soon getting himself hired at Detroit's WKNR FM and then WABX FM. But he'd get back to Boston ... eventually.

In addition to describing a circle of skinny-dipping, joint-smoking film-makers around a college swimming hole discussing the artistic merits of filming an orgy, the *Boston Globe* story also mentioned that four WBCN disc jockeys attended the conference. Norm Winer was swept up by the same carousing spirit that Parenteau witnessed: "This is where Atlantic Records signed J. Geils Band on the spot. ... Dr. John was there ... Baba Ram Dass [the spiritualist] and Jerry Rubin. It was the first time we met many of our counterparts. They shared our philosophies and convictions; it really fortified and energized us. We weren't just crackpots clinging on to an unrealistic goal—there were other people sharing that." Andy Beaubien drove to the conference with Charles and Norm: "It was strange and bizarre, but fun and exciting. I remember driving back, all energized, but also politicized. We came to the station and we all went on the air and had a discussion, kind of a debriefing, taking phone calls live. There was this sense that this was the beginning of a major change."

"We were very utopian in our way of thinking," Jim Parry acknowledged.

"We wanted to be crazy, committed, but responsible, music-loving human beings," Sam Kopper stressed. "Those were our ideals and makeup, and I'm proud of that."

"A lot of the decisions were made by all of us for a long time," added Al Perry. "There were occasionally some interesting arguments, but I think we stood for the community."

It was, as the song said, the dawning of Aquarius, and a spirit of unity bonded the members of 'BCN's young staff, inspiring them to reach out to serve their brotherhood of listeners. Kate Curran, who came to the station as "an indentured servant for Charles," as she jokingly referred to her unpaid status, headed up the effort to establish a daily schedule of volunteers who would be available to answer listener calls. "Charles, kind of, put me in charge. At first we had people answering the business phone [because] there wasn't a separate line. If it was a business call, they'd hand it over to the secretary. It was very confusing. Then there was a separate line, but we'd have issues if a volunteer didn't show up for a shift because the new phone line would just ring and ring and ring. The secretary didn't have time to answer it, so she'd just wrap it up in a blanket and put it in a bottom drawer." Those early efforts, though, resulted in a formal WBCN Listener Line that took requests and provided information about the songs being played, upcoming concerts, a list of available rides to different cities, the lost cat-and-dog report, and answers to a myriad of questions. "We hated to have to say, 'I don't know.' Curran and her second-in-command, Arlene Brahm, obtained reference books, including an encyclopedia set, to answer even the most random questions a listener could venture. "The Listener Line grew exponentially," Curran added, and soon the air staff counted on it to be there. "Danny Schechter looked at the volunteers as his own little slave pool," she remembered with amusement. "His thing was, 'Oh good, they have food!'" Of course, as Bill Lichtenstein could attest, sometimes being in the right place at the right time on the WBCN Listener Line could be a very good thing indeed.

David L. Bieber in the June 1970 *Boston* article wrote, "A further means of activating the WBCN audience is via public service announcements, jointly handled by Andy Beaubien, who does the afternoon show, and [J.J.] Jackson. 'We can't put on a Robert Goulet Heart Fund appeal because it doesn't relate to our audience,' says Beaubien, 'but we can perform a real public service by giving out the information about the presence of Project Place (a sanctuary for down-and-out young Bostonians) or the need for volunteers for the Cambridge Free School." Listener Line volunteers often referred callers to these organizations; those on bum acid trips were most often advised to phone Project Place if they needed more help (and the phone hadn't melted in their hands). The article also mentioned some WBCN-produced public interest programming that had been presented on the air, including Laquidara's "Eco-Catastrophe," a twenty-five-minute documentary about the environment, and a special prepared by the women's liberation organization Bread and Roses. Aired on International Women's Day, that program targeted male prejudice in the media and trumpeted the group's battle for equality in the workplace.

WBCN's involvement with Bread and Roses resulted from a local firestorm sparked by Charles Laquidara after he recorded a public service announcement for Project Place's Drug Dependency Treatment Center. Sam Kopper, who caught most of the flak as program director, recalled, "They were looking for more doctors and office staff. Charles did this spot where he said, 'If you're a guy, we need doctors; if you're a chick, we need secretaries.'" The Valentines' Day 1970 edition of the *Boston Globe* identified the quote as, "If you're a chick and you can type, they need typists." Nevertheless, the article also reported that the response to this slip of chauvinism was immediate (and creative): "About 35 young women, protesting 'male supremist [*sic*] policies' at the hip rock music station, WBCN-FM, swarmed into the station's Stuart Street studios yesterday and threw eight live baby chicks on the station manager's desk." One of the protesters, according to the article, made the group's position clear to Kopper: "These are chicks—I am a woman." The *Boston Globe* went on to interview Debbie Ullman, one of four females on staff at the time, who allied herself with Bread and Roses: "WBCN isn't taking a leadership role in the women's liberation movement." She then pointed out that the male members of the staff were distressed by the action: "It's the first revolutionary issue in which they've been confronted as the enemy." Laquidara defended himself by saying that the needs of the Drug Dependency Treatment Center were immediate and the announcement recorded quickly. However, the gaffe still exposed an embarrassing bias at the station, even if the male staffers didn't realize they had it.

The infamous baby chick incident gave Kopper real clout with Ray Riepen and Mitch Hastings to support the concept of incorporating a greater female presence on the air. Debbie Ullman was considered for a full-time shift, but after a summer of fill-in work, she flew the coop. "I was under the spell of Woodstock and had the urge for going; I was ready to get into somebody's car and drive to California with my dog." The West Coast sojourn didn't last: Ullman would soon be persuaded to return ("I was living in a cow field with an artist in Mendocino") and take over the morning show. But in her absence, the search continued to find another hip female jock with radio experience, and eventually Kopper heard good things about twenty-year-old Maxanne Sartori at KLOL-FM in Seattle. He contacted the DJ and asked for a tape of her work. "I loved the way she sounded right away." Maxanne made her way east, sliding into the afternoon two-to-six slot on Friday the 13th of November 1970, just as the station's beloved J.J. Jackson accepted an offer to do radio at KLOS-FM in Los Angeles. While many lamented the loss of one who had already become legendary, the community also accepted and embraced Maxanne's rejuvenating enthusiasm and energy. It is not fair to say that her star at WBCN would eclipse even J.J. Jackson's but rather that Maxanne's contribution to the radio collective would take the station to an exciting new level.

Big changes were also afoot for overnight jock Norm Winer, who left WBCN for a new job as program director at CKGM-FM in Montreal. It was a "sign of the times" job offer, as Winer recalled: "The station owner flew to India several times a year to meet with his guru, and he hired a whole bunch of spiritual young Canadians to run the radio station. [Then] their PD took a leave to take up with [his] guru for two years, so they needed [someone] to fill in." Exit Winer (in a flourish of sitars) for his promising new gig. He'd be back in six months, barely long enough to be missed, but in that time he hired Sam Kopper away, leaving Charles Laquidara in charge of the WBCN air staff. When a sudden managerial shakeup at CKGM tossed Winer back to Boston, he immediately sought out his former boss: "I came back from Canada, it was March or April 1971, and I remember following Ray Riepen around his apartment, trying to talk him into letting me be program director. I said I had a plan and that I had been [a] PD for six months and knew how to read the ratings, or something." Since sparks had been flying between Laquidara and Riepen, Winer's pitch worked. "He'd had enough of Charles. So, I came back to 'BCN, after being the rookie, to being the boss, to people who weren't really used to having a boss. Then, [the goal was to] have a cohesive plan that we all could agree on, creating some minimal rules to have a consistency so we could continue to create and program collectively. That was the strength of the station, well through the seventies anyway." Kopper also limped back to WBCN after his even briefer Canadian career of a mere six weeks, but his old morning shift had already been doled out to the recently returned Debbie Ullman. After all of these musical chairs had finally been positioned, the feminist listeners in WBCN's audience had to have been pleased: by the spring of 1971, a year after the Bread and Roses intervention, WBCN now had a female announcer in both the morning and the afternoon shifts.

Maxanne, as she simply referred to herself on the radio, felt grateful that Laquidara had set off the chain of events that made her attractive to hire, even though Sam Kopper maintains that it was talent that landed her the job. She joked to the *Boston Globe* years later in 1983, "You could say I owe my career to Charles because he's the one who said *chicks* on the radio." Bill Spurlin, who worked at WBCN as an engineer at the time, observed in

Maxanne loved to rock, punished WBCN's studio speakers, broke Aerosmith, and introduced Bruce Springsteen. With Kenny Greenblatt in the Prudential Studios. Photo by Dan Beach.

his blog, "Max was a very striking, handsome woman, [and] a woman on the radio was a New Phenomenon in 1971. The idea that a woman could control that stream of power was enough to shake my male-centric, woman-in-the-kitchen prejudices, and reinforced the very power of the medium itself." It was quickly evident that Maxanne liked to rock, as Debbie Ullman observed: "I was motivated by the counterculture—played Jesse Colin Young, Incredible String Band, Holy Modal Rounders, BB King, Jefferson Airplane, Velvet Underground; [but] she was really into the rock and roll. She was much more tuned into what [would be] happening with 'BCN by the later seventies."

"A lot of what she played are the songs you hear on Classic Rock radio today," added Tommy Hadges. Maxanne would eventually be credited with championing Boston groups like the J. Geils Band, Aerosmith, and the Cars, and counted some less famous bands from the area as favorites, including Reddy Teddy, Nervous Eaters, Fox Pass, and Willie "Loco" Alexander. Hadges continued, "She was the one that was really able to find the cuts that seemed to resonate with the audience." In regard to "resonating," Maxanne also liked to *feel* the music she played. "She ran those speakers at the loudest possible level imaginable," Hadges joked seriously. "The production room at Stuart Street was right next to the air studio; Andy Beaubien and I would try to get some work [done] in the afternoon. But sometimes it was difficult because Maxanne had her speakers up so loud that the whole place was rattling!"

With its latest lineup in place, the on-air collective headed into 1971 as darlings of the underground media, standard-bearers for the countercul-ture bohemia entrenched in Boston, and heroes to so many young bands. Artists were offered free reign to drop in at Stuart Street for interviews and to play live on the air. "I remember one of the first live broadcasts was Jesse Colin Young and the Youngbloods," Kate Curran recalled. "They set them up front in the sales office and they played for an hour or two. Dr. John came in and all the volunteers and hangers-on were there in the Listener Line area going, 'Wow, it's Dr. John!' On his way out he looked at us and said, 'Catch you in the moonbeams.' It was soooo Dr. John!" Others as diverse as Hound Dog Taylor, Van Morrison, Pete Townshend, and Gary Burton also dropped in. One historic evening in November 1970, Laquidara hosted an acoustic guitar summit on his show featuring Jerry Garcia, Bob Weir, and Duane Allman (who had less than a year to live).

To further the egalitarian goals of a hippie "revolution," yet still bring in the cash needed to keep business going, the staff needed to confront and compromise with the typical business model that made any radio station's survival possible. The most radical viewpoint was to not sell advertising at all, like college radio stations. But where would the money come from with no school or foundation underwriting the programming? The WBCN compromise was to place a certain limit on the number of commercials allowed on the air, a practice begun by Ray Riepen and agreed to by the staff. "One of the initial deals was eight ads an hour," the general manager said. Because demand usually outpaced inventory for the first twenty-four months after the change from classical music, this choice worked for the station. *Boston* magazine reported that as of May 1970, the rate charged for a one-time, one-minute commercial in AAA time (the most coveted positions in the evening) on WBCN was $32, and the net sales for that March "rose to the highest in the station's history, surpassing all other FM stations in the city." The 20 May 1970 edition of the *New York Times*, in a story entitled "Around Country, FM Turns to Rock," pictured Laquidara working in the studio and pointed out that WBCN's shift from classical to rock had "enriched the station by $41,000 a month."

The WBCN collective considered the style and presentation of the commercials to be equally important to the number. Some accounts, like any of the U.S. Armed Services or the tobacco companies, were flat-out refused. Other national companies that might be welcomed were asked to leave their slick and professional, agency-produced commercials at the door, and ads with jingles in them were also turned down. A potential advertiser had to agree that its spots could be rewritten, produced, and voiced by 'BCN announcers, an attempt to adapt the messages to echo the "underground" attitude at the station and create palatable, even entertaining, vignettes for its listeners. Tim Montgomery, who replaced Debbie Ullman in sales in 1971, recalled, "Up until 1973, maybe '74, we didn't run any prerecorded commercials. It didn't matter who the advertiser was; they were all either [done] live or we produced them." With his rate cards and station information stored, ironically, in an old army surplus gas mask bag, Montgomery canvassed local businesses looking for advertisers. Then, "I'd sell the ad, convince them that WBCN was the right kind of radio, but then I had the task of saying, 'Well, you might have a lovely spot, but we can't run it. [But], believe me, we know how to talk to our listeners.'" Montgomery quickly became one of 'BCN's most successful rising stars. "I must have written thousands of ads in the time I was there. Then I had to voice and produce them myself. I had no experience doing that; I just picked it up." Montgomery became "the voice" for Underground Camera, a major supporter of the early WBCN. "They were ahead of their time, in thinking that you should use a regular person to voice the ads. Of course, everybody on 'BCN was an ordinary person, I suppose, in their delivery and voice. There were no actors behind the mike at WBCN." Laughing, he added, "Except for Charles!"

Constant debate over which commercials would be acceptable to WBCN's hip, young audience led to disagreements between the staff and Ray Riepen, who turned bitter remembering that aspect of his 'BCN stewardship. "I got screwed a few times because they, literally, turned down [accounts] instead of taking them and telling them what to do [to conform with 'BCN's image]. Kinney Shoes and the big companies were calling me; Coca-Cola called up, and the guys in the sales room would not take them."

"No one wanted Coca-Cola on WBCN because it epitomized 'The Man,'" Tim Montgomery stated. Nevertheless, some in sales felt that, with the proper tweaking, the soft drink giant's message might be acceptable. Some sample Coca-Cola commercials were mailed up from the ad agency in New York so the DJs might find some creative ways to rework the content. In the meantime, Tim Montgomery, as the new sales manager, headed to Manhattan as part of his first national sales call. "My first time in New York; I was probably all of twenty-five years old and a little green. My [national sales] rep and I went to McCann-Erickson, which was the agency for Coca-Cola." The powerful advertising firm had been a key player in developing "I'd Love to Teach the World to Sing (In Perfect Harmony)" into a massive 1972 Coke marketing campaign. But that jingle, also a best-selling Top 10 hit in America, clashed hideously with 'BCN's radical musical views and countercultural slant, and should have portended the debacle to come.

Montgomery headed into his big meeting with McCann's media buyer. "I introduced myself. She was a fairly burly woman behind her desk. She stood up, looked me in the eye, and said, 'Who the fuck do you think you are!' I was, sort of, blown back into my chair; I'm going, 'Uh . . . Uh . . .' This was my greeting, my first call!" Montgomery sat there stunned. The customary civility and decorum associated with any high-level business negotiation had been rudely tossed out the window—and flown back in his face. He was totally mystified as to why. The woman held up a small cardboard box for Montgomery to see; it was one of the mailers that agencies used to ship reel-to-reel tapes of commercials. This particular box looked like it had been mailed to WBCN and then returned. "I looked at it and I saw that someone had written in grease pencil on it, 'FUCK YOU!' Whoever was in production at the time, I think it might have been Sam, got the tape at the station and mailed it back like that! Suffice it to say, it was a long time before we got any Coca-Cola ads!" When asked about his possible role in the matter, Sam Kopper only smiled deviously and pled ignorance.

The inmates ran the asylum, but Riepen didn't want to believe it. Debate over the commercials extended into the programming. As WBCN had been set up to be a true free-form music vehicle, these disagreements were seen by the staff as sacrilegious. A famous story in this regard began with a normal Sunday midday show hosted by John Brodey, who had only recently conquered the stage fright he'd had on his initial shows. "I couldn't speak; my voice just froze. I'd just play music—for hours." Brodey was finally getting

John Brodey, digging a Ginger Baker drum solo. Photo by Dan Beach.

his rhythm, and as Charles Laquidara put it in an essay he wrote in 1978, "Sunday radio was a snap: just throw on a little Joni Mitchell, and if you need to organize, put on a side or two of *Woodstock*." Laquidara would probably need some of that organizational time since he was relieving Brodey at five and had spent the entire day tripping on acid: "Riding the cat, falling in the park, sitting paranoid in the shade, listening to 'Let It Be' and 'Bridge Over Troubled Water' on the stereo, drawing pictures, philosophizing, and finding easy solutions to problems that had stumped the world's greatest minds." Still buzzing, Laquidara drove in while listening to the station. "Good old John Brodey was playing the Ginger Baker cut from the Blind Faith album. Thump-thump tha-thump. 'What a mothafuckin' drummer!' I thought."

"You could play anything, and I took that pretty literally," recalled Brodey, who isn't sure exactly what he was playing that day: "I think it was 'Toad' because the live version does have a twelve-minute drum solo." So, there's a discrepancy about whether the cut was "Do What You Like" by Blind Faith or "Toad" from Cream. Which one it was is not necessarily important, since both featured English drummer Ginger Baker performing one of his trademark solos for a very *long* time.

"The red phone in the studio blinks; it's the hotline . . . somebody important," Brodey recalled. "It was Ray, who immediately says, 'What the fuck are you doing?' I'm like, 'Oh my God, it's the big guy!' He said, 'What in the hell made you think that playing a ninety-minute drum solo on a Sunday afternoon was a good idea?' I said, 'Well, I . . .' Then he said, 'Shut up!' And then something along the lines of, 'That's going to be your last drum solo!' I thought, 'That's it, I'm done.'" Riepen admitted, "I never called [the jocks on the air], but I did that day. I said, 'Hey pal, that's a tune-out! You can't play a drum solo for twenty-two minutes regardless of what kind of station we've got!'" By the time Laquidara walked into the studio "Brodey looked pale. He was visibly shaken. He explained that the station president had just called him on the hotline and blasted him for playing a long drum solo." Laquidara was outraged at Riepen's action, which in his view violated one of WBCN's fundamental constructs. "'Oh John, I'm sorry, he's not supposed to do that. He should know better.' John went home totally deflated and there I was, all alone in that studio, my head still spinning from breakfast"—LSD breakfast, no less.

Some strange and uniquely personal gears clicked in Laquidara's mind at that point, setting him on a course that most would not have taken. From his 1978 essay, he said, "I put the long version of 'Toad' on the turntable and turned on the mike. 'Good afternoon. This is WBCN in Boston. My name is Charles Laquidara. We have this boss who thinks he has impeccable taste, and he sometimes likes to impress his friends so he calls up the announcers on the hotline and makes requests or gives orders, or yells at us. It's hard to do a good show after the boss calls, and poor John Brodey, the guy who was just on before me, got this call from our boss and was yelled at because he played a drum solo. I guess we should settle this once and for all.'" Laquidara admitted being plagued by self-doubt as his inner voice of reason fought for control: "God, Charles, wait! What the fuck are you saying? People out there must think you're crazy." Nevertheless, outrage at Riepen, who had now become "the Man," and definitely some residual chemical agents, prompted Laquidara to ignore his unease and push the turntable switch. Ginger Baker took off on his extended rhythmic romp; and that was just the beginning. The DJ rooted through the vast WBCN library and came up next with "Mutiny," a seven-and-a-half-minute drum piece from the Aynsley Dunbar Retaliation, following that with some classic jazz thumping from Buddy Rich. "When they found me two hours later,

I was wrapped around a beanbag chair in a corner of the air studio. Side 4 of *Tommy* was just ending. 'It's okay, Charles. Man, that was beautiful. The whole town's talking about it. You're a hero!'" But he was certain he'd be fired, as he wrote in his essay: "A hero? An unemployed hero. A hero on welfare. You-got-any-spare-change-mister hero. I can't even work at the *Phoenix*. The son of a bitch owns that too!"

"I've got nothing against Charles," Riepen protested, "but he has always tried to give this image himself of being this really hip guy who always had to fight to create his terribly important show on WBCN. I mean, Charles Laquidara wasn't the morning king when he worked for me; he was just another disc jockey. I probably said five things to him in the whole time [we] were there." In regard to the drum solo incident, Riepen said, "I might have had discussions with some of the guys at times, but I didn't care what they played. I never was on anybody's back." Nevertheless, things quickly went from bad to worse when Riepen brought in Arnie Ginsburg to act as WBCN's general manager. As the legendary AM radio king who had ruled the Boston airwaves for ten years, Ginsburg was not necessarily a bad choice for boss since he knew the many aspects of radio business and engineering intimately. But because he embodied the hyperactive, pimple-cream-selling, screaming and "teening" AM Top 40 style—the very antithesis of what WBCN represented—this appointment was viewed by the staff as an enormous slap in the face and a potential threat. "When Arnie Ginsburg came in we were all flipped out. Here was the guy that was the enemy," Sam Kopper remembered. "We were fearful Ray and Arnie would tighten us up and turn us into Top 40."

Desperate for information about Arnie Ginsburg's plans, and paranoid enough to do something about it, several members of the staff hatched a scheme. "That was the time that Ray felt threatened by Charles," Norm Winer noted. "There was concern to make sure we knew what was going on."

"We bugged his office," Kopper revealed gleefully. "We took a microphone and ran a cable above the ceiling from Arnie's office to the engineering room."

"The engineer was Mike Ward," Winer added, "who also used to do the occasional Sunday morning shift until we realized he was just tracking Savoy Brown album sides and playing boogies at 7:00 a.m. But, he was a great hippie engineer. From [his] office in the opposite side of the building, we could hear every meeting, phone call, everything going on in Arnie Ginsburg's office."

"I set it up, but it was not all my idea; we had meetings and voted on things," Mike Ward recalled. "It was a full-size broadcast microphone on a stand sitting up in his hung ceiling, pointing down through a vent screen [over] his desk. If you looked close, you could see the chrome gleam." Nevertheless, Ginsburg never had cause to glance upward, and the apparatus remained undiscovered, "for weeks, possibly months," Winer figured. This small example of pre-Watergate espionage much resembled some of the government's underhanded activities that 'BCN so often railed against, but in the interests of the "revolution," the bugging was seen as justified.

Arnie Ginsburg tightened up the rules around the place, but his main concern was transferring and upgrading WBCN's transmitter from the (old) Hancock Building to the top of the Prudential Tower (the new Hancock Tower, then rising on the Boston skyline, was significantly impairing the station's signal). While this went on, the staff knew his every move, even being forewarned and given time to scatter when the general manager decided to inspect the premises. Ward remembered, "He was one to interrupt a reefer break in the production studio, causing us to hastily vacate via the back staircase. Then, casually wrecked, [we'd] come back up the elevator," like nothing happened. Ward attached a tape deck to the mike and was able to record anything the conspirators wished. Kopper laughed, "But we were idiots, the Three Stooges of espionage!"

"One day the cleaning lady left a chair on his desk while doing the floor and did not return it," Ward elaborated. Ginsburg came in to work the next morning and was greeted by the unusual sight. "So, being suspicious, he looked up from the chair."

"The sun, perhaps, glinted off the metal," Winer supplied. "So, he followed the wire and tracked it down to the engineer's office . . ."

". . . to the tape recorder," Ward finished. "I got word from a secretary out in front just in time to remove the tape . . . I denied everything." The engineer didn't get fired for the incident, probably because he had a lot of complicated transmitter work to accomplish for Ginsburg, but as Ward pointed out, "Let's just say I was on Woo-Woo's list to replace!"

With Ginsburg threatening to fire conspirators in the wake of Stuart-gate and the frequent skirmishes with Laquidara over the content of his show (he once asked everyone in Boston to flush their toilets at the same time, a request the DJ had to rescind once too many people phoned in to warn him that the system couldn't handle it), the mood at the station seesawed between uncertainty and outright paranoia. Laquidara was actually fired by both Riepen and Ginsburg but quickly rehired because of staff dissent and listener response. The members of the WBCN collective discussed ways it could protect itself, and the best way to do that, according to Danny Schechter, who had studied at the School of Industry and Labor Relations at Cornell, was to form a union. He continued, "Instead of a normal broadcast union like AFTRA (American Federation of Television and Radio Artists), which is just for on-air personalities, we wanted to find [one] where everybody could be a member, as part of the whole spirit of 'BCN. So I researched it and found the United Electrical, Radio and Machine Workers of America union (UE), which had been a progressive [labor organization] in GE plants. It was sort of left wing and had 'radio' in the title, even though they were referring to manufacturing radios." The UE accepted their new shop, and after the 'BCN staff took a "yes" vote and signed the official papers, Danny Schechter was elected first shop steward. At the time, the employees at WBCN had no way of knowing that this union would play a historic part in the station's survival just eight years later.

The establishment of WBCN's union was the final straw for Riepen, the station's largest shareholder. "It broke my heart that [Danny Schechter], who I gave a job to, organized my station. What could they complain about? It was just a deal to be hip for them." The power struggle just wasn't worth it to Riepen, who had also been dealing with similar staff disagreements at the Boston Tea Party and the *Phoenix*. Those disputes had resulted in the hippie entrepreneur selling his interest in the concert hall in late 1970 and bailing out of the newspaper the following May. Don Law, who picked up the remains of the Tea Party, marveled, "It was amazing that [the Tea Party, Phoenix, and WBCN] all unraveled at the same time. I don't think any of us saw that coming."

"The reason I sold 'BCN was that they stabbed me in the back," Riepen remembered with sadness. "I walked away from it." The *Boston Globe* reported in its 12 September 1971 edition that he sold his interest for "for $238,000 and some change" and "officially extricated himself from WBCN-FM, where he had been persona non grata ever since the station's staff had rebelled over what it thought was his 'artistic interference' in running the station."

Within days Riepen would vanish from a scene that he had significantly influenced.

WBCN's yearlong power struggle was unknown to most of the station's listeners, who could still call up the Listener Line and report a missing pet, request "Room to Move" by John Mayall, or ask one of the eager staffers how many ounces made up a "key." Joe Rogers returned to the station in '72 after some disheartening experiences with the underground radio scene in California. The somewhat cynical prodigal son told James Isaacs at the *Real Paper* that WBCN featured "the most tasteful radio in the country. You'll hear less shitty stuff on it than anywhere else." The station pushed the boundaries with Jimmy Byrd's gospel program Sunday through Friday from 6:00 to 7:00 a.m. and Little Walter's oldies show on Sunday nights. Starting in 1972, Eric Jackson, widely referred to years later as the "Dean of Boston jazz radio," spent five years at WBCN. The station began broadcasting many shows from the newly opened Intermedia Sound recording studio on Newbury Street, including Jerry Garcia and Howard Wales, the New York Rock Ensemble, and Loggins and Messina. Winer recalled that "Jim Messina, from Buffalo Springfield and then Poco, was a good friend of Charles. His album with Kenny Loggins had been out on Columbia for six months and it did nothing . . . just sat there on the shelf. We had them come in and do a broadcast with some people watching on a Friday night and it was incredible! Literally, the record took off [after that]. They sold out of it for multiple weeks at the Harvard Coop. It really got the project going."

A similar Intermedia broadcast in March 1972 featured Canned Heat on a snowy winter's night. "We all got stuck inside," Winer laughed, "so what was supposed to be an hourlong show turned into multiple reels of tape. Some members of the J. Geils Band had nothing to do and were hanging around, [so] Peter Wolf sat in. The singer, Bob 'The Bear' Hite, assumed that Peter Wolf, being the singer of a well-known band, was J. Geils. So he kept calling him Jay all night and Peter never corrected him!"

"Plus, he kept saying Jay 'Jiels,' not Giels," laughed Charles Laquidara. WBCN also instituted a live broadcast series from the Jazz Workshop nightclub on Tuesday nights, thanks to the engineering prowess of Sam Kopper, who was now heavily involved in the production of live performances for the station. Winer remembered broadcasting such icons as Chick Corea, Miles Davis, Herbie Hancock, Stan Getz, and Rahsaan Roland Kirk, although nothing much, if anything, remains of that series on tape. "In those

Broadcast of Canned Heat and friends from Intermedia Sound, March 1972.
Photo by Charles Daniels.

days," Kopper remembered unhappily, "there were shows I didn't even record because I had so little money. I couldn't afford the tape!"

With Riepen gone, Mitch Hastings reclaimed the reins at Stuart Street. But in the aftermath of a debilitating brain operation, the owner's already-peculiar personality had taken some strange detours. WBCN engineer Bill Spurlin described Hastings like this: "I glimpsed T. Mitch moving rather feebly in and out of his shabby office. He seemed very, very old to me. His mostly bald skull was marked by two shallow depressions about an inch across, surrounded by liver spots, the marks, I was told, of a more-or-less successful brain surgery. Wearing sports jackets that appeared to have been styled in an earlier era, hunched, bespectacled, we could occasionally hear him in his office [talking] in a high whining voice." Nevertheless, no matter what debilitated state Hastings might have been in, he still managed to accomplish a great deal. The WBCN transmitter was relocated to the top of the fifty-two-story Prudential Tower, eliminating most of the signal issues, and work began to move the entire station into the same prestigious Back

Bay facility. When Arnie Ginsburg bailed in 1972, Hastings promoted his sales guru and sometimes-jock Al Perry, to replace him. The choice delighted the staff since "Crazy Al" was one of their own, experiencing the many seismic changes at the station from even before the switch to rock. Perry remembered, "Mitch was back in the picture and probably figured, 'Whatever I want, Al will go along [with],' which wasn't necessarily the case."

Charles Laquidara followed his muse on the air, wherever that took him, but since his show was quite popular, he became virtually untouchable. Despite a few instances of being disciplined and fired, he always returned within days. A particular piece of underground radio folklore, one of his infamous best, emerged from this time. It's a psychedelic fable, and the names may have been changed to protect against flashbacks. "Randy was this beautiful woman and Robert was her boyfriend," Laquidara remembered. "They lived in this rather famous place in Boston on Storrow Drive where they have the sign 'If you lived here, you'd be home now.' They invited me to come over. Robert said, 'Randy makes the best matzo ball soup—ever!'" Charles would never turn down a meal, a fact corroborated by Kopper: "His appetite blew me away. I couldn't believe how much he ate and still stayed rail thin." Along with a friend, Laquidara planned to drop in on the way to do his nightly radio show. "We had some good grass, so we smoked a joint on the way over, parked, and then went upstairs." Laquidara sat and talked with the couple, eating and breaking out the party favors in return. But even though the smoking continued, Laquidara kept his eyes on the clock, eventually calling an end to the visit so he could get over to WBCN. "In the meantime, Randy kept bringing out this soup and it was so good that I had three bowls of it."

"We left in my car and I was really high. I said, 'Man, that was really good smoke!'" Laquidara chattered abstractedly in marijuana-speak while driving across Back Bay, tossing out ideas for his radio show while sitting at the stoplights. Once he arrived at the station, though, it was time to focus and get organized for the shift.

> I went to the record library and pulled a bunch of albums, got my turntables all cued up and the carts with all the commercials or special 'BCN songs ready. The program started smoothly, but then, about a half hour into the show, I went to make a segue and noticed there was no record on the turntable. I said, "Shit!" I asked whoever was there, "Get me a record, anything!" While

that was happening, I put the mike on: "WBCN . . . Charles Laquidara . . . what a beautiful night . . ." I began to read a commercial, and as I started, things began getting a little weird. I said, "You know, I'm reading this commercial, but I have to tell you that all the words are blending together . . . kind of like the words are . . . melting. So, I can't finish this commercial right now . . . but I will play a song and come back later and finish." So I started the record, whatever it was, shut off the mike and went, "That is *really* strong shit you got, man!" My buddy goes, "My shit is not *that* strong!" And then we both looked at each other and suddenly I knew: Randy and Robert! *"She laced the soup!"*

At the moment, Laquidara was keeping it together, but he was drifting higher with every passing second, blubbering details of the evening live on the air, even though, due to FCC paranoia, he didn't quite admit he had voluntarily gotten high. "You know, we might have had some weed, but I don't know what happened. We were over at our friends' place, and they gave us Randy's matzo ball soup that Robert said was really good and I'm positive she put mescaline in the soup! So, now I'm having some trouble on the radio and I'm going to call someone because I cannot do this alone. We all know the rules here, you know . . . I know it's far out that she makes good mescaline soup, but you're supposed to tell people!" A random thought poked its way into his consciousness and fought to be noticed until finally finding voice on the air: "What I'd like to do is meet everybody under the Citgo sign [in Kenmore Square] at midnight. So we'll talk about this and maybe Randy and Robert will be there to explain. I'll just hang in here as long as I can and play whatever I can come up with."

For a time, Laquidara made a good fight of it, resisting the mental taffy his brain was twisting out. "All I could do was play carts with commercials [on them] from a rack by the [broadcast] board. I couldn't possibly get up, select an album in the library, and put that delicate needle on a certain spot on the record. I mean, I was definitely going to fuck that up, so I didn't go near the records!" The DJ played commercial after commercial after commercial, sometimes getting a song that just happened to be on a cartridge. Finally, "Someone came in and said, 'Okay Charles, you're okay. We'll take over now.' Laquidara managed to compose himself a bit and did, indeed, head over to Kenmore Square. "We went over to the Citgo sign at midnight and there were 250 people there, cheering me on! It got all

Charles Laquidara on three hits of matzo ball soup.
Photo by Sam Kopper.

muddy after that, so I don't remember the rest . . . but that's what tripping was about . . . you went on a voyage."

As the zaniness and candor continued to unfold every day on 104.1, the front office remained busy, too. In June 1973, Mitch Hastings realized his dream to move WBCN out of its funky environs on Stuart Street to literally enter the clouds, occupying plush new digs on the fiftieth floor of Boston's Prudential Tower. The move heralded another era for the station, still motivated by its hippie dream of community-based integrity, but now boxed in by the mainstream-business perceptions created by stepping up to a high-rent "$200,000 studio," as reported by the *Boston Sunday Globe* a month before the move. Al Perry opined, "It wasn't conducive for us at all. In retrospect, we probably should have been in a three-family [house] in Brighton or the South End. But that was his vision . . . and his vision was always just to sell the place. Ray was gone and Mitch was back in charge and he had the board of directors where he wanted them. I could see him selling them that bill of goods: 'We're going to move the station to the Pru; yes we're going to spend a couple hundred grand to get there, but its going to come back twentyfold when I sell it.'"

"It didn't seem like the right thing to do," Joe Rogers remembered. "We didn't belong there, and we were aware that this was part of the plan to make 'BCN more marketable."

"We had mixed feelings," Andy Beaubien added. "Stuart Street was really familiar to us . . . it was home, and the overhead was really low. Mitch rationalized it by saying we'd have the transmitter and the studios in one place, but it still didn't overcome the fact that our operating expenses greatly increased."

Taking up one half of the Prudential's fiftieth floor, the new facility was described by John Brodey as "a little fishbowl" because the building observation deck ran along WBCN's outside wall. Large windows offered Skywalk visitors a direct view into the daily workings of a radio station if, of course, they chose to look away from the impressive views of Boston in the opposite direction. Sam Kopper scratched his head at the move: "It was such a weird place to be. Why would you want this progressive rock radio station inside a window with all of Middle America to walk by?"

"It was a little creepy for me," Joe Rogers laughed. "It always felt like someone was watching me over my shoulder." Despite the oddities of being in this new, alien place, the staff, at least, grew to enjoy the amenities of the facility. Even Rogers had to admit, "They *were* better studios and new equipment."

"When we moved to the Prudential," Kate Curran added, "Everyone on the Listener Line was bemoaning the integrity [of the station], and the sellout, but all I thought was, 'Oh good! It will be clean!'"

"Nobody minded being in a place where there weren't holes in the wall and you had the luxury of a big air studio with the library right there in the same room," Brodey added. "Mitch Hastings walked me through and showed me everything," including the five-layer-thick glass walls to deaden sound, wall-to-wall carpeted floors "floating" on pads, and a lead shield between the station and the roof. "We all wanted to go up on the roof and take a look [where all the antennas were], so an engineer took us up. He said, 'Grab that fluorescent tube.' We went on the roof and held that [tube] up in the air and all of a sudden it lit up—all by itself! [from the radio frequency radiation streaming out of the antennae]. I thought to myself, 'This can't be good for you!' Then we weren't laughing about that lead shielding. Hastings was a genius."

Perhaps so, but the man with a marvel of engineering talent also had questionable skills as a business leader. Five years into WBCN's mission,

Hastings made a high-risk maneuver, placing the station underneath a mountain of debt even as it sat throne-like atop the Boston skyline. While fattening the calf was clearly the intention, could the dream that Ray Riepen had originally fostered survive in such an environment of intense expectation and demand? Even if they could endure under Hastings's new order, what would happen when the absentminded boss sold his creation? Would the new owners care about the magic of a Joe Rogers segue, an Allman Brother jamming on the air with his friends in the Grateful Dead, or Maxanne introducing a young local outfit named Aerosmith? Would they find the airtime for Danny Schechter interviewing a clandestine FBI informant, someone on the Listener Line talking a late-night caller out of suicide, or Laquidara's latest escapade. What about "Lockup"—targeting Boston's prison population—or the special programs championing women's rights, the dangers of pollution, racial equality, and gay tolerance? But even if the staff felt that sitting on the pinnacle of a skyscraper was inappropriate, WBCN had still become one of Boston's most successful FM signals. So, perhaps this lofty new environment in the clouds was not completely out of character, even for a formerly ragtag bunch of radio hippies.

CAMELOT

Ensconced in their soaring steel and glass castle in the clouds, the scruffy inhabitants of Boston's hippie radio station now found themselves under-dressed as they arrived at the skyscraper on Boylston Street. There was also the uncomfortable check-in at the security desk before every trip up the high-speed elevator. Flashing the ID became routine, but signing in various and sundry guests in their assorted states of coherency could be awkward. After a rush-inducing rocket ride up to fifty, visitors stepped out onto a floor shared by WBCN's main entrance and the popular Boston Skywalk, with its panoramic view of the city and surrounding land and sea. Instead of heading toward the inevitable flocks of tourists, station regulars could angle to the unmarked side of the building and slip around a corner through the discreet rear entry. As far as the building management was concerned, that was fine, because the less they saw of WBCN's employees, the better.

Early on in the station's tenancy, T. Mitchell Hastings approached Norm Winer to discuss the matter.

> He said that he had heard from the people in the Prudential Tower that sometimes our attire was a little uneven, that some of the DJs actually looked a little sloppy, disheveled, and unkempt. So he said, "I was thinking that we should have blazers with a station logo on them [made] for everyone, and it would be very nice."
>
> I said, "Well, Mr. Hastings, that's an interesting idea, but how about something a little less formal. How about a jump suit with people's names sewn on them? Both men and women could wear them."
>
> "Hmmm, interesting." He wandered off with that thought, and it never came up again. That was a big challenge of those years: just trying to distract him [or] change the subject so it would slip his mind.

Debbie Ullman had departed again, and Buffalo radio veteran Dinah Vaprin was brought in to replace her in the mornings. Vaprin remembered, "Charles would build the morning show into the empire it became, but during my years it wasn't like that. We thought no one listened to rock and roll radio then; they got up at noon and went to sleep at five o'clock in the morning." Vaprin's personality had an edge. "I was a flaming radical feminist type," she laughed. "I could make people uncomfortable." At one of the announcers' meetings, Norm Winer asked if anyone had questions. Laquidara recalled, "Dinah pipes up, 'Yeah, I've got a question. We're supposed to be so hip and cool and yet, you have women working the shit shift in the morning.'"

"Yeah, I probably said 'shit shift,'" Vaprin admitted. "I said I was tired of getting up at four o'clock in the morning!"

"I'm at the back of the room," Laquidara continued, "and I say smarmily, 'Dinah, there's no such thing as a shit shift on this station. You should appreciate any shift!' She yells back, 'You try getting up at four in the morning and have nobody listening to you!'" Laquidara rose to the bait: "Okay, I don't care; I'll take it!"

"Charles was good for a dare," Vaprin said. "He took [the shift]—and the rest, of course, is history."

As morning DJ, Laquidara introduced a new term to Boston radio: "The Big Mattress," his medley of comedy bits, social satire, political commentary, wake-up calls, and music. "It was a hippie thing," he told *Virtually*

Alternative magazine in 1998. "Everybody's waking up; we're all brothers and sisters on this big mattress. There was nothing sacred; we'd make fun of the Pope, all the Boston icons, the mayor, the president. We were calling the White House and getting the Secret Service tracing our calls, and the FBI coming in—it was pretty heavy shit. The show just caught on because it was so different and so unique and so sacrilegious." Although it sounds like Laquidara embraced the up-tempo Top 40 radio style that WBCN had fought against for five years, he really lampooned that approach. "Charles was just doing this off the wall parody of AM radio, for stoners," summed up his morning editor Steve Crowley, aka, "Mono," who assembled stories for the show's regular news reports. Plus, Laquidara certainly never committed the sin of featuring only the top "forty" records in the country. For him, as zany as the presentation and talk became, the music was always a key player: "We could do so much with music; you could grab everybody by their balls by playing The Doors' 'Five to One' and in the background the sound of some kids getting shot at Kent State."

A cast of characters formed around the morning show. Sixteen-year-old Tom Couch, who volunteered at the station answering phones and would later grow into the role of WBCN's production director, saw a slip of paper Laquidara had tacked up in the Listener Line booth one day. "[It] said he was looking for a production assistant that he could pay ten dollars a week to get coffee, help with things, and maybe learn some production," Couch recalled, laughing. "People had written all over the note, things like 'Charles, you're a slave trader!' 'How can you pay someone as little as ten dollars?' 'What are you, "The Man"?' But I said, 'I'll do it!'" A young temp-agency veteran who had done stints as a substitute teacher and dental clinic worker, who went by the singular moniker of Oedipus, wandered in one day to become a similar jack of all trades for Laquidara—at worst picking up coffee and at best doing some writing for the show. Michael Fremer, a Boston University law student who had a natural knack for comedy, had been creating commercials for Music City, a local record store. "Charles liked the voices I was doing," Fremer recalled, so he became a regular contributor, performing impersonations and eventually producing "three- or four-minute political cartoon bits" entitled "Can I Have My Money Back?" which ran on the show for years.

Another character who developed on the earliest days of the show was "The Cosmic Muffin," whose daily astrological reports began after

Darrell Martinie: "It's a wise person who rules the stars, and a fool who is ruled by them." Photo by Don Sanford.

Laquidara made a comment on the air, announcing something akin to the moon being in Leo-Virgo. Darrell Martinie, a devoted listener, instantly phoned to correct the DJ. "The moon can be in the sign Leo or in the sign Virgo, but not both at the same time," he patiently explained. From that initial encounter it was determined that the Milton native had indeed studied the stars and knew what he was talking about. It was proposed that Martinie would write and record daily astrological reports, Laquidara dubbing him "The Cosmic Muffin," taken from a line in National Lampoon's album *Deteriorata*, which declared, "Make peace with your God, be he hairy thunderer or cosmic muffin!" Martinie jumped at the chance, and the timing was perfect: he told the *Boston Globe* in 1979 that at the time, he had no job and only $38 to his name. From a daily feature on "The Big Mattress" to three reports a day on WBCN, and eventually a regular feature on a network of radio stations, Martinie gauged his astrological predictions like a weatherman forecasting imminent climactic conditions. In 1974, Tommy Hadges suggested rating the day with a number, for example, giving it a "4" for going out or a "9" for staying in. Although Martinie initially resisted, the practice would later give the reports their most distinctive highlight—that plus the famous disclaimer at the end that he discovered in an astrology book: "It's a wise person who rules the stars, and a fool who is ruled by them." "A requirement of the FCC," he told the *Globe*. "Otherwise I'd have to say, 'The preceding has been brought to you for your entertainment.'" How bland. And Darrell Martinie was always anything but bland.

As August 1973 arrived, with the paint still drying on the fiftieth floor, Norm Winer decided he had to lay off Dinah Vaprin. "At this point, I don't even remember why," he mused. "She was very influenced by the women's liberation [movement], but she wasn't [let go] because of her politicization."

"I got fired, or laid off, whatever. There was clearly a financial issue, and I think I probably had the least strong ratings. A lot of the reason they hired me was because I was a woman, not because I was deeply knowledgeable about rock and roll music. I would play a lot of blues, a little jazz, and women artists; this didn't really fit in with WBCN programming, even back then." In any case, Vaprin's departure made big waves once the word got out, and Norm Winer and Al Perry soon faced an unforeseen public relations firestorm. As in the chicks-on-the-desk incident from three years earlier, local women's liberation activists quickly fanned the sparks of protest into a major conflagration. "I didn't really have that much of an audience, not like Jim Parry or Charles, but the women that responded [to the firing] were a smaller, very noisy section of the listening public. They reacted, the union got involved, and it became, for lack of a better word in the seventies, viral!" Vaprin marveled.

"Firing Dinah when I did led to some massive demonstrations with people accusing me that I had fired her because she was a woman," Norm Winer recalled. "I explained to them that, in fact, it was because she was a woman that I hadn't fired her until then. It was overdue."

People didn't see it that way. The United Electrical, Radio and Machine Workers of America union reacted, drawing in shop steward Danny Schechter and Charles Laquidara, who joined the effort to rehire Vaprin. Fellow employee and radical feminist Marsha Steinberg, who worked at 'BCN under the moniker of "Jamaica Plain Jane," threw in her support as well. The union circulated amongst the station employees a petition, which announcers, news personnel, sales, and office staffers all signed. The petition stated in part, "Dinah's removal has denied WBCN's listening community of a valuable source of new musical perspectives at a time when commercial broadcasting seems to be moving farther and farther from responding to community needs. There are only two women on the air at WBCN. We cannot afford to lose her voice, her presence, her tastes." After the story hit the *Real Paper* and *Boston Globe*, an ad hoc organization that called itself the People United for Free Air Waves called for a boycott of any products advertised on the station, magnifying the personnel dispute into a larger political arena. In a flyer entitled "Put Feminism Back on the Radio," the group asserted, "WBCN cannot hide behind its slick and hip rock 'n' roll cover any longer. Dinah's removal is an overt act that exposes WBCN for what it is—a big-time business. WBCN is more interested in soliciting and accepting anything that is marketable than in providing the air time for different cultural communities. For example, they have been freely advertising Portuguese made wine, such as Costa de Sol, which many people are boycotting because of Portugal's massacre of Africans in Mozambique."

It was clear that Vaprin's firing had focused the ire of many who saw the innocence and dreams of the sixties being replaced by the strident rebound of a mainstream consumer attitude—as exemplified in the evolution of their beloved local radio station. This wasn't just about a woman's rights; it became a blow against freedom, a signpost marking the threatening return of an empire. Depending on a WBCN listener's views, they could end up anywhere on a curve from righteous believer to growing cynic, and in this period where apathy was not a problem, it seemed like everyone had an opinion . . . and expressed it. Another unidentified, more militant protest flyer from August 1973 stated, "We are sick of the sham of the 'hip rock radical groovy' FM station. Your air waves no longer sing or speak to us, or for us." The underground, radical radio station of old, which, in reality was still as liberal as any commercial station in the early seventies, was being reimaged by its opponents as a compromised entity: a tool of Nixon's America or, at least, the Portuguese wine industry.

Those opponents would always be there, replaced in time by others who had different axes to grind, but to Winer and Perry's credit, the issues stirred up by Vaprin's firing were taken seriously and addressed. A statement was drafted and presented to the radio station from representatives of the "Boston Women's Community," a grassroots group that made several demands of WBCN management to resolve the "Dinah Vaprin Question." The document called for a minimum of two full-time women announcers to always be maintained; the reinstatement of Vaprin to the payroll; the creation of a two-hour women's public affairs show by 1 October 1973 with

Vaprin as producer; and the introduction of layoff protection language in the station's union contract. Whereas this was not a legal document and didn't need to be treated as such, the radio station responded by creating a brand-new women's program. "We produced this show from 6:00 to 8:00 p.m. on Monday nights," Vaprin recalled with pride. "It was a cultural show, not a political one. We were just trying to give women a voice, or a platform from which to be heard. It was the first women's programming on the radio in the city." Although she wasn't restored to her full-time announcer's position, Vaprin worked hard on the show until 1975, when she left, albeit voluntarily.

Meanwhile, some others were on their way out, with a bang in Michael Fremer's case. On the air as a weekend announcer mixing comedy and music, he soon ran afoul of the feminists. "I used to joke about Marsha Steinberg, 'Jamaica Plain Jane,' who worked in the news section with Andrew Kopkind. I called her 'Jamaica Lame Brain.'" For International Women's Day on 8 March 1974, WBCN planned a twenty-four-hour cycle of special women's programs and music ending at midnight, which happened to be the beginning of Fremer's weekend show. At the stroke of twelve, he launched into his introduction: "Welcome to the 'Men's Room!' Tonight we're going to play men's music: like Herbie Mann and Manfred Mann, the Johnny Mann Singers . . . *Man*drake Memorial! We're going to have the 'Shaving Hotline' where you can call and get grooming tips and learn how to grow that funny little thing that Gregg Allman has under his lip! We've got a list of words that are not acceptable, like menopause and mandate." Fremer went on and on for "ten minutes of silliness." The feminists were not amused, raising enough of a fuss that Norm Winer had to fire the comedian. Fremer had *man*aged to *man*euver himself permanently off the air as a DJ at WBCN.

The *Boston Globe* reported in 1973 that 'BCN produced "an ambitious array" of public affairs programs, including "'Community Report,' 'Third World Report,' 'School Daze,' and 'Lockup,' the reigning favorite at Boston area prisons." Listeners behind bars tuned in to hear their requests as well as self-help information for them and their families. 'Lockup' had been in production for a couple of years by that point, and Al Perry remembered an encounter from 1971 regarding the show. "We needed to raise money to buy radios for the prisoners who didn't have them. A lot of guys were writing in, saying, 'I've got to listen to a guy's radio three cells down.' Frank

Rockin' at the Government Center with Jonathan Richman. WBCN voter registration rally in front of Sam Kopper's Crab Louie broadcast bus. Photo by Sam Kopper.

Ron Wood and Rod Stewart of the *Faces* bring star power to the voter registration rally. Photo by Sam Kopper.

Zappa's '200 Motels' was coming out and Frank was doing a movie tour. I guess because I lived close by, they said, 'Al, you got to get Frank; Maxanne's going to do an interview with him. He's at the Elliot Hotel; pick him up and bring him to the station. Then you can talk to him about helping [us to raise money].'" The assignment was not necessarily an easy one.

I'd met him before and Frank could be touch and go. But he finished the interview, we got in the cab, and we're going back to the hotel and I say, "Frank, I've got a favor to ask."

He says, "Yeah? What?"

"The owner of the theater [where the Boston premier of '200 Motels' was scheduled] is a friend of ours, and we have this prisoner request show . . ."

"Really?"

So, I told him about the show, then, "We'd like to buy some radios with all the ticket sales at the premier." Frank was a straight-shooter. Right away, he looks at me and says, "No problem." Writes on a piece of paper, "Here's who to call." And that was it.

Then there was "The Lavender Hour." John Scagliotti tells the story of that groundbreaking gay-oriented show.

Sunday night, ratings-wise, was quiet, and we in the news department wanted to be DJs once in awhile, too. So, we convinced Al Perry to let us do a thing in the ten o'clock hour called "Pot Luck." You could do anything, it didn't matter, and anyone could do one of these shows. Eventually, Andy [Kopkind] and I just took it over and we changed the name to "The Lavender Hour." It allowed us to play music, but then we started to add things, like John Wieners doing these gay poems. When David Bowie came to town, I went out to the press conference with a tape recorder and asked him why he wouldn't come out! [Bowie had told the English press that he was bisexual, admitting years later that it was just a publicity stunt.] I had a big fight with him, which I taped and then put on the air.

Scagliotti laughs at the memory.

Danny Schechter quickly welcomed and sanctioned what he called the "innovative gay cultural magazine" as part of the station's public affairs schedule. Scagliotti clarified,

It wasn't like a gay documentary; it was just music, poetry, and once in a while an actuality or a sound bite. But it turned out to be the first time a gay program was on commercial radio. We got to play disco music which nobody would play on WBCN. Can you imagine 'BCN playing Donna Summer? I started getting letters from kids way up in New Hampshire and Western Massachusetts. One sixteen-year-old said that he had thought of committing suicide because he was gay, but would sneak up to his bedroom and listen as quietly as he could to "The Lavender Hour" and it helped him through his crisis. It never occurred to me that we had gone from just having a good time to being important to people.

In 1972, the station was the most open place around; you could be gay and be loved and cared for and allowed to work. You could do anything you wanted to, to the point where I was real spoiled. I had a hard time working in other places after that.

"Very few stations inspired that kind of loyalty, commitment, and devotion to the art of it and also the audience," Danny Schechter observed. "This was beyond the rock and roll and the format; you were part of something, and it made going to work something you looked forward to doing. We really felt it was *our* radio station and we were acting, in a broader way, for the entire community."

If WBCN was a big fractious family at the dinner table, someone still had to bring the roast out, say grace, put out fires (in the oven *or* at the table), deal with the owners of the house and reset the smoke detectors. That person was Norm Winer, who as program director realized that if he ever wanted to do a normal air shift like the others, he'd need to develop an alter ego. "I decided I needed a different identity so people wouldn't apply for jobs or critique the station when I was on the air. I became Old Saxophone Joe. I stole that name from Bob Dylan's song 'Country Pie,' but I tried to convince people when I met them that he wrote the song about me!" Truly, Winer was a laid-back music junkie, just wanting to be left free to enjoy his radio show, unshackled from the responsibility of the office while on the air. To some extent he succeeded in segregating his roles, but any time, day or night, the program director might be thrust into the role of cop or peacemaker. The constant parrying with Hastings; the episodes with Vaprin and Fremer; the need to address new trends in music; and increasing pressure from sales to relax the standards that the air staff had

WBCN's Norm Winer as Old Saxophone Joe, with
Sam Kopper. Photo by Don Sanford.

imposed on both the type and style of commercials allowed made him less
'BCN's communal shaman and more an official administrator. "I believed
that unless I assumed this responsibility, no one else would, because none
of us wanted responsibility."

John B. Wood at the *Boston Globe* wrote an article in May 1973 about the
station's appeal and, at the time, the impending move to the Prudential
Tower just a month hence. Entitled "WBCN: Somewhere between Artistic
Freedom and Anarchy," the piece detailed the "creative chaos" that the
station's twenty-three employees experienced daily while running the successful radio outlet and maintaining its hip attitude, "on the edge" music
choices, and cool restraint. But Wood isolated WBCN's advertising policy as
"one bone of contention" and described it as "an unfortunate corollary to

its success." The article continued, "In its idealistic early days, the station rejected any ad that was inconsistent with its programming or its scruples. Those days are gone. Today, WBCN takes any advertising it can get. National advertising, which accounted for less than $100,000 last year, is expected to make up half of BCN's $800,000 billing this year." The article mentioned Winer himself, on the air as Old Saxophone Joe, lamenting that he had to play a taped ad "extolling the praises of Exxon gasoline." WBCN had made a commercial shift in its thinking, but the process was gradual and marked by pitched battles back and forth. Tim Montgomery, then sales manager, observed, "Norm was always really laid-back and soft-spoken, logical. [But] we had a lot of arguments over the advertising because I was always trying to wedge spots on the air. I tried to force the issue because we did have budgets to make; T. Mitchell always saying, 'I got to make the payroll.' So, there was that tension."

"I remember the list of unacceptable advertising that we had, in Charles's handwriting, actually," recalled Winer. "Dow Chemical, who made napalm for the U.S. Army was on that list. We didn't have to run all of that because we were lucky we had a booming advertising base with all of Don Law's clubs and concert halls, the head shops and waterbed stores. I remember thinking back in the early seventies, 'If it wasn't for waterbed stores, 'BCN would have gone under a long time ago!'"

The decision as to whether an ad was acceptable, even if it made it through the Winer/Montgomery negotiation process, could often be personal and emotional. The best example involved, who else, but Charles Laquidara, one morning on "The Big Mattress." "You have to remember the context," Montgomery explained as he recounted the story. "The Vietnam War was still raging; it was unpopular, and you did anything you could to stay out of it. People my age had either moved to Canada, dodged the draft, or become a Peace Corps volunteer like me." Montgomery worked for Vista, the domestic version of the Peace Corps, gaining a selective service deferment before working at WBCN. "One of my big accounts was a store called Underground Camera. I was driving into work one day, Charles was on, and he started reading one of my ads." Norm Winer, already in his office, recalled that "Charles was doing a live sixty-second read about 'Honeywell-Pentax Week,' the big discount sale at Underground Camera in Cambridge." Tim Montgomery resumed the story: "I'm tooling along in my car, kind of not paying too much attention, and he does the spot pretty straight. Then he pauses and says, 'Well, that's fine if you want to support a company that's responsible for killing babies in Cambodia.' What!? I had a hard time keeping my grip on the steering wheel after that!"

Danny Schechter attributed Laquidara's ad-lib to a report the News Dissector had read regarding Honeywell's production of antipersonnel weapons for the U.S. Defense Department, including the deadly cluster bomb known as the "Lazy Dog." These bombs didn't explode themselves but scattered hundreds of round canisters, each with its own delay fuse, over the landscape. Later, these bomblets would release hundreds of deadly projectiles, either ball bearings or steel darts, indiscriminately through hapless soldiers or innocent civilians after the people had emerged from hiding. Honeywell's association with Pentax Cameras and, in turn, the local retailer, created a set of advertising bedfellows that Laquidara castigated in one quick coda. "Before Charles was even off the air, the general manager had marched down to my office; the client had already called and the salesperson was in there," Norm Winer said. "I told them, 'I'll take care of it!' So, very quietly, I congratulated Charles on doing a great thing, then publicly suspended him with pay till it blew over. We could tell the client, 'We disciplined him; he's off the air.' But the reality was that Charles did the right thing. That was part of the romance of 'BCN: doing the right thing." There wasn't much Tim Montgomery could do: "That was the end of the account," he moaned, "plus, they sued 'BCN."

The case went to trial and John Taylor ("Ike") Williams, the lawyer who defended Laquidara, clarified that there were two claims: one for defamation, which he considered a weak case, and a far stronger one regarding a breach of the advertising contract. "Charles was very difficult to prep because his attention span was about ten seconds. He didn't seem to realize that it could be a really serious case. Then, he shows up late, he's got a jacket, but no tie, and he's got a Great Dane on a leash which is causing havoc in the courtroom. He couldn't remember any of our prep work; he was pretty much a nightmare witness!" When the court calmed down, Laquidara took the stand. "I told Charles, this lawyer will play the super-patriot and you'll be Mr. Hippie Draft Evader, that's how the scenario will look to the judge, so don't play into that; just say yes or no. Of course, there weren't two questions before Charles is getting into it with the lawyer." The result seemed to be a major blow for the defense, but Williams's second, and final, witness, who happened to be an expert on armaments, more than made up for the DJ's damaging episode. He described Honeywell's "Lazy Dog" bomb in detail, shocking the judge, "who didn't like it at all," Williams added. When the judge returned with his decision, "He found that it was true: Honeywell was the manufacturer of the Pentax Camera and also bombs that killed many civilians in Cambodia, including babies. So, no defamation; [it was] substantial truth." However, he did find that Laquidara had violated his duty of fair dealing with Underground Camera, possibly damaging the advertiser. But, to the dismay of the plaintiff, the imposed punishment was purely symbolic, as Williams summarized: "He awarded them one dollar; the damages didn't cover one minute of their lawyer's time."

Laquidara's behavior wasn't always inspired by a noble and ethical purpose. Tim Montgomery recalled, "The Australian Wine Board heard about the legendary WBCN, hired an agency and wanted to run some ads. They also wanted to bring someone from the board and an ambassador or consul from Australia to be interviewed by Charles. They were going to spend a lot of money, by our standards; so I took a deep breath, went to Charles and said, 'I beg of you, please be on your best behavior.' Well, that was my first mistake, asking him to be cool with these people." Montgomery chuckled before continuing:

> The whole group came in for a live interview: two people from the agency, two people from the consulate and two from the wine board. Everything was going great; then, sure enough, in the middle of things, Charles goes, "Hey, you guys ever heard this?" And he played the [now] famous Monty Python comedy bit about Australian wines. So, you've got these [six] stone-faced people sitting in the studio looking at Charles as he broadcasted [a comedy sketch] about Australian wine "with the bouquet of an Aborigine's armpit." No one got the joke; the agency was ripped! Suffice it to say, we lost that buy too.

Norm Winer sheepishly accepted a small amount of responsibility for Laquidara's Monty Python fixation. When he had returned to WBCN in '71, the new program director brought the first album from that British comedy troupe home with him from Montreal, where *Monty Python's Flying Circus* was a popular show on Canadian television, but would not air in America for another three years. "WBCN was famous for playing bits off of every comedy and spoken word album of the time," he recalled. "Joe Rogers, in particular, had memorized all of them: Fireside Theater, Congress of Wonders, Conception Corporation and the Credibility Gap. We'd play a spoken word bit between songs; it was a punch line that would add another level to the audio entertainment." The DJs loved the Monty Python record, and listeners seemed to embrace the bizarre humor as well, even though most critics seemed baffled by it. When the British troupe completed its first movie, *And Now for Something Completely Different*, Winer contacted the management at the Orson Welles Theater in Cambridge to arrange a late-night, WBCN-only screening. "We talked about it on the air and that night the crowd stood around the block." This gave Monty Python one of its original footholds in America, eventually helping to lead the oddball group to massive success and, even, cultural permanence. Winer related that, later on, the station sponsored a petition drive to prompt WGBH to pick up the television show, "which led to us getting the screening for *Monty Python and the Holy Grail* [in 1975]. We gave coconuts to each of the BCN listeners who attended: to duplicate the sounds of the horses galloping [in the movie]."

WBCN enjoyed a long relationship with the Orson Welles, which specialized in independent, art, and foreign film releases. Perhaps the most famous association between the station and the theater was in bringing the 1972 Jamaican classic *The Harder They Come* to the city. Starring musician Jimmy Cliff, this movie sowed the seeds of a burgeoning island music style known as reggae, which had captivated England in the previous few years but still hadn't made much of an impression in America. John Brodey became 'BCN's "initial ambassador," as he put it, since he vacationed in Jamaica every year and had fallen in love with the beats of the island. "When I was down there, I went to an open-air theater and saw *The Harder They Come*; I thought it was amazing. That led to a screening and, after that, midnight showings—for years." Perry Henzell, who wrote and directed the groundbreaking film and compiled its influential soundtrack, mentioned in a 2001 interview in Jamaica, "It had the second or third longest run in American cinema history in Boston at the Orson Welles Theater. It played in this one theater for seven years straight." The movie showed viewers the unvarnished reality of the Trenchtown slums but also introduced the novel reggae beat and some of the musicians that played it, including Jimmy Cliff, the Maytals, and Desmond Dekker. Norm Winer recalled, "When Jimmy Cliff finally came to Boston to play, we arranged to bring him over to the Orson Welles." As the midnight showing of *The Harder They Come* was ending, Winer and friends led the star quietly through a door and into the darkness of the theater. "Then, when the lights came up, Jimmy Cliff was standing there in person, wearing that T-shirt with the star in it that he wore throughout the film. It blew people's minds! It was so damn cool."

"We played every significant artist in the reggae world; it was a vital part of the 'BCN legacy," Winer emphasized. "For several years we were sure it was going to be the next big musical movement."

"Jimmy Cliff and all these reggae artists saw that this was where it was happening," recalled John Scagliotti. "They saw us playing it and people coming to the clubs to see them, so Boston became this, sort of, center of reggae." Bob Marley and the Wailers chose the city as a jumping-off point for the 1973 tour, their first extended stateside visit, to support the U.S. release of *Catch a Fire* on Island Records. Fred Taylor, the booking agent, manager, and eventual owner of Paul's Mall and the Jazz Workshop on Boylston Street, shook his head in amazement remembering those five nights. "It was the original band with Bob Marley, Peter Tosh, and Bunny Wailer. Reggae had taken off because of *The Harder They Come* at the theater, and I paid attention to the music because of that. Kenny Greenblatt at 'BCN, who was a gem and one hell of a marketer, had convinced the film company that they should not use the traditional method of print advertising, but, because it was all about the music, put the promotion money into the station."

Norm Winer remembered meeting the Rastafarian musicians for the first time: "We were all excited, so we got there in the afternoon when they were setting up. As a peace offering, we brought some stuff to share, to break the ice. We didn't realize that for Marley, [our joints] were like hors d'oeuvres, so it just never came back around in our little circle. I guess he figured it wasn't full-sized, it must be just for him." Sam Kopper was also astounded at the considerable consumption of the Wailers, who continually showed the Boston hippies how it was done.

I parked my broadcast bus in the alley behind Paul's Mall and started pulling cables out. On every horizontal surface just outside the stage door, there were multiple roaches just laying around. We're talking Jamaican-styled roaches that were so full of ganja that you could roll two American joints out of one

Bob Marley roach. When the band showed up for sound check and when they started playing—they started smoking! Paul's Mall had a ceiling only about seven feet high, and in very short order, you couldn't see from one side of the room to the other. I spent the whole afternoon running around in that . . . air.

Kopper, though, probably wasn't "running" very quickly, nor in a straight line.

Just across Boylston Street in the WBCN studios at the Prudential Tower, John Scagliotti waited to conduct an interview with the band. "Bob Marley and the Wailers were walking around, hanging out. Most people didn't know who these musicians were yet, so nobody was really impressed, but I was just ga-ga. I thought it was the greatest thing that they were there. They came into the production room, which was the size of a little closet, and there was Bob and, I think, three Wailers, so it was pretty crowded. As soon as they sat down, all of them pulled out these huge spliffs, like cigars, and they lit 'em up." The newsman realized with a shock that anyone strolling by on the Prudential Skywalk could gaze right in through the picture windows, so he whirled around and whipped the curtains closed. "Then the room became filled, I mean, *completely filled*, with smoke. There was no way the air system could take it out fast enough. They didn't even offer me a hit, but I didn't need it! I was totally stoned." Scagliotti did the interview and obtained quite a scoop, considering how famous and important Marley would become in the future. "I brought them out of the room, and I tried to introduce them to people, but I was falling all over the place, out of my mind. They left and I came back in the studio [to listen to the playback] and I realized *I hadn't hit the record button!* I can laugh about it now. It was the best interview—*ever*, but you'll just have to believe me."

Even as critics questioned the wisdom of moving the hippie radio commune into the ivory tower of the "Pru," the station thrived in the midst of a most inspired phase. The breadth of music heard on the air rivaled that of any FM rock station of the time, and the many live concert broadcasts the station presented from 1971 to 1975 demonstrated that diversity. There was blues from Canned Heat and the reggae of Bob Marley, while former Frank Zappa band member Lowell George and his new outfit Little Feat brought the funk/rock. Fred Taylor remembered others: "We did Randy Newman, and a Dr. Hook & the Medicine Show broadcast from Paul's Mall that almost put WBCN off the air. Ray [Sawyer] was the main vocal-

WBCN visits James Taylor and Carly Simon at their house in 1976. Note: that's Tommy Hadges stylin' in the white pants. Courtesy of the Sam Kopper Archives.

ist; he had the patch over one eye, and he used the f-word like it was a conjunction!"

Then there was the R & B–infused rock and roll poetry of a young singer/songwriter/guitarist out of New Jersey. Bruce Springsteen was a barely known entity outside of Asbury Park, but his determined gigging up and down the East Coast had slowly built a small following. Once his debut album, *Greetings from Asbury Park, NJ,* came out in January 1973, Springsteen played Boston relentlessly. Just a few days after the album release, he was in town for a full week of concerts at Joe's Place and Paul's Mall, stopping in at WBCN with members of his E Street Band to do an interview and unplugged performance on Maxanne's show. The assembly played six songs including "Blinded by the Light" from his album and a cover of Duke Ellington's "Satin Doll," utilizing guitar, accordion, and even a tuba. Although Maxanne never mentioned this interview as being a peak moment in her career, Springsteen's eventual superstardom and the impassioned legion inspired by his music and lyrics would fondly elevate this amiable visit to iconic status. However, a year later, Springsteen's second rendezvous on Maxanne's show would be the one recorded and "bootlegged" on vinyl and

CD with such frequency that it became one of the most highly sought-after broadcasts ever conducted. Even today, the April 1974 appearance is readily accessible from Internet sites. The visit, much the same as the first, utilized players from the E Street Band in a rollicking and informal acoustic performance. Once again the assembly did a half dozen songs including "Fourth of July, Asbury Park (Sandy)" and "Rosalita" from Springsteen's second album *The Wild, The Innocent, and the E Street Shuffle*. The unique and hilarious performance of the latter is easily one of the most memorable nine minutes in WBCN's entire history.

Then there was hard rock, best exemplified by a scruffy and pimply local band named Aerosmith, which released its first album the same day Bruce Springsteen released his. "The first person ever to play our record was Maxanne, who was on 'BCN in the afternoons," Steven Tyler mentioned in the band's biography *Walk This Way*. Maxanne involved herself deeply with the Boston music scene, loved the hard bluesy rock of Aerosmith, and persistently championed the group to Norm Winer and the other jocks. Nevertheless, everyone missed the appeal of a band that channeled the Yardbirds, James Brown, and the Jeff Beck Group. "I refused to let Maxanne play Aerosmith in the beginning," Winer confessed, "I thought they were too derivative. But, of course, she was right." The program director decided to let 'BCN's listeners be judge and jury, and then admitted defeat when all the positive reaction on the phones just blew him away. Winer even green-lighted a live broadcast of the band from Paul's Mall on 23 April 1973. That show, which featured songs from the debut album, plus some covers, showed the WBCN staff what Maxanne had sensed all along: this was a terrific live act and one that the station should invest in. Some of the folkies on the air staff might not have liked Aerosmith all that much, but any group that took a serious crack at "Mother Popcorn" by James Brown had to deserve some respect.

In August 1975, WBCN's first serious challenge from another radio station abruptly shouldered its way onto the FM radio dial. Previously known for playing innocuous elevator music, WCOZ began broadcasting an automated selection of album-rock favorites from The Who, the Eagles, and Rolling Stones among others. In short order, the owners at Blair Radio hired some talented radio personalities to fill out the schedule. These new challengers emulated the format and personas established by 'BCN's veterans, speaking conversationally and concentrating on younger lifestyle issues, but sharply reducing the variety of the music selections played. Not unlike the AM Top 40 approach, though not as extreme, the station installed a format that whittled down the available library to just the most appealing tracks, as determined by requests and market research. Andy Beaubien, who would later work as a DJ and program director at the new station, commented about its arrival: "We didn't take 'COZ very seriously. To be perfectly honest, we had a superior attitude, which was endemic at 'BCN. We were kind of elitist: 'They're nothing compared to us; we're the originals, we're the best.'"

"We were certainly aware of what they were doing, but we felt it was most important to perpetuate what it was that 'BCN represented," Tommy Hadges observed. "Maybe we were naive in the beginning; we didn't know how much impact ['COZ] would eventually have in the marketplace."

Indifference to the new challenger soon led to a grudging awareness. Almost immediately, WCOZ had an impact on the veteran station as listeners who might not enjoy the many colors of the musical palette suddenly had an alternative to switch to. A rock music fan who would have remained listening to WBCN while Old Saxophone Joe segued from the Stones into a Sonny Rollins saxophone piece could now flip the dial to find "Won't Get Fooled Again" on 'COZ. It was such a dramatic incursion that by mid-1977, WBCN's Arbitron rating stood at 1.7, while its chief tormentor had stomped to a 4.6. "The downhill thing for me was the whole 'COZ thing and the ratings that they got," bemoaned Al Perry. "Maybe we needed to clean up our act a little bit, but traditionally, in the history of radio it happens a lot: someone comes along, beats the big guy up for awhile, the big guy makes adjustments, stays the course and people come back." But the general manager was constantly kept under pressure by Mitch Hastings, who, despite his frail appearance and absent-minded behavior, became surprisingly irritated and tenacious. "Mitch wasn't a stand still kind of guy. He kept saying, 'There's something wrong, there's something wrong.' I said, 'We just have to stay the course.' But, he didn't want to stay the course; he kept the pressure up until finally I said, 'I can't do this, it's too much.'"

Al Perry's departure marked a sea change for WBCN. Then there was another serious blow: Norm Winer, who had presided over the circle of DJs for five years, also decided to move on. "I was personally disenchanted with some of the stuff [going on] in Boston, given the ideals of the time. The school bussing thing really disturbed me; Boston had been a horribly racist place in previous generations, and we thought we had escaped that.

My girlfriend of many years split and my home in Concord was empty with me being the only one living there. That was all part of my general disenchantment." When Winer received an offer to replace the morning man at KSAN-FM in San Francisco, he pounced. But, as critical as were the double departures of Perry and Winer, fate hammered a third stake into WBCN's heart in '76 when Charles Laquidara completed his final show and quietly skipped town. Particularly hard to accept, by those who knew the truth, was that the host of "The Big Mattress" was not jumping ship to another radio station but leaving to embrace the new love of his life: "I was heavily into cocaine, and the show was getting in my way," Laquidara remembered matter-of-factly. The official excuse at the time was that the DJ was leaving to pursue his passion for acting full time. "Yeah, that's because I couldn't tell everyone I wanted to go off and do cocaine for the rest of my life. But, when I left, I literally told friends, 'I know this [is] going to kill me, but it's such a great way to die!' That's how much I loved it." When Laquidara came in for his last broadcast, only a few people had a clue that it was the end. He had prerecorded the last two hours of the "Mattress" the day before. "When I started the show, I announced to everybody I was leaving. I did the first couple hours live, then I started the tape, and just walked out. I was through with radio for the rest of my life, and I was on my way to Vermont."

As Laquidara drove west on Route 2, further and further from the station that had made him famous, away from a stunned group of coworkers and thousands of surprised listeners, he focused with delight on his new, if impulsive, freedom. Seduced by the taste of a new relationship, an obsession that would most likely kill him, the ex-morning announcer bopped blissfully along, tugging on a big, fat joint and listening to the car radio the whole way out to the northern interstate, even as the fringes of static began to grow and grow. Then, as the roach cooled in the ashtray and the car whizzed past the pines along Route 91 north, no amount of fiddling with the knobs on the radio could prevent the atmospheric noise from completely crowding out any trace of 104.1 in Boston. But by that point, Laquidara wasn't even trying.

THE BATTLE JOINED

It's interesting to consider that one of WBCN's cornerstone talents, who would remain with the station for thirteen years, developed his Boston radio career by working to co-opt the very audience that "The American Revolution" had fostered since 1968. Ken Shelton would participate in a near-dismantling of Camelot, a surprise attack from across the radio dial that WBCN's denizens, absorbed in their own greatness, would remain badly naive of until it was almost too late. As one of the standard-bearers at the upstart WCOZ, Ken Shelton was destined to plunder large numbers of listeners from the ranks of a station once considered bulletproof, protected by the unwavering support from its community since inception. But as the seventies passed the halfway point, 'BCN, in a languished state with a depleted rank of veterans, found itself vulnerable to a new rival with roots in an old approach—the Top 40 system of concentrating more attention on fewer songs.

As a kid in Brooklyn, Ken Shelton grew up loving the music of Elvis Presley and the pop hits of the late fifties and early sixties, furtively listening at night to a small transistor radio hidden under his pillow. Later, he discovered and embraced the sixties counterculture, hanging out in the Village and seeing the Fugs and the Blues Project as well as the Band at the Fillmore. He graduated from college with a bachelor's in speech and theater, ostensibly for a career in television. "It was 1969; Vietnam was still going on, so the goal was to stay in school," he remembered. "I looked around for graduate schools and ended up at BU." When Shelton arrived in Boston, the first thing he did, like everybody else, was set up his stereo system. "I scanned up and down the dial and suddenly I heard 'Witchi-Tai-To' by Everything Is Everything. [I] never heard that before. So, for the next two years, I did not move that radio dial from 'BCN; I couldn't tell you the name of any other station."

After graduation, considering Shelton's distinctive deep voice and passion for radio, it seemed a likely career choice, but marrying his college sweetheart meant that he better bring home the bacon—and quickly. Since he'd studied for a job in television, Shelton got a job at Channel 4 (WBZ-TV) as an assistant director, working nights, weekends, and holidays, and becoming the floor manager on Rex Trailer's *Boomtown* kids show. "Rex used to ride his horse into the studio, and one time, there was a big [accident] on camera," Shelton recalled, pinching his nose and chuckling. "Somewhere was a pot of gold and some creativity in TV production, but it sure wasn't there!" Then, the restless director ran into Clark Smidt, one of his grad school buddies, who had worked his way into WBZ's radio division selling ads for the AM station and managing the younger FM signal, which broadcast classical music by day and jazz at night. Shelton expressed his unhappiness, mentioning he was tired of sweeping up after . . . the talent. Smidt rapidly became Shelton's ticket, finagling his buddy into the radio division as his programming assistant.

When the parent company, Westinghouse, decided to flip WBZ-FM over to a contemporary Top 40 rock format, the station became an automated jukebox with Smidt and Shelton picking the songs and voice-tracking the entire broadcast day on tape. "It went on the air December 30, 1971, and the first record was *American Pie*, the long version," Smidt recalled. "It was a tight playlist, but we did slip in a few album cuts." The pair came up with innovative features like the "Bummer of the Week," where listeners suggested a song they hated, and the DJs would, literally, break the vinyl on the air. The format quickly generated impressive ratings, even if the company didn't want to put too much effort into it. "The general manager came in one day," Shelton recalled, "and said, 'We don't want you guys to have too much of a presence; it's the music [that matters], so from now on, no last names on the air. You're Clark. You're Ken.' I had a little desk in Clark's office and he would call over to me, 'Hey Captain!' I thought that 'captain' had a nice ring to it. They said no last names, but nothing about nicknames, so, I started using 'Captain Ken' on the air and it stuck like Krazy Glue!"

At the end of '74, Smidt and Shelton made a fervent pitch to Westing-house for more resources, but the company balked at the idea of pumping additional dollars into WBZ-FM. "The head of Westinghouse thought FM was just a fad," Shelton bemoaned. "Our hearts were broken. Clark didn't stay much longer, and I was only there a few months after that." But the lessons learned in that experience proved that there was room for a station in Boston that played a mixture of rock-oriented Top 40 hits and appealing songs from LPs, which had become the medium of choice for much of the college-aged audience. In 1972, Blair Radio had purchased WCOZ, a beautiful music FM station with the slogan of "The Cozy Sound, 94.5 in Boston." Three years later, "Clark got the job as program director to turn 'COZ into a rocker and go against 'BCN," Shelton stated. "I was the first person he hired."

WCOZ switched to rock programming on 15 August 1975 and then, during the Labor Day weekend, aired a syndicated special called "Fantasy Park: A Concert of the Mind." Smidt marveled, "What a reaction that got!" Originally produced at KNUS-FM in Dallas and broadcast in nearly two hundred markets, the two-day "concert" generated excitement in nearly every market it aired. Songs from existing live albums were assembled to make it appear as if all the best rock bands in the world were together on the bill of some incredible concert that was being broadcast as it happened. "It was a three-day fictional Woodstock-like event designed for radio," said Mark Parenteau, soon to be hired by the newly minted WCOZ. "People weren't sure if the concert was real." David Bieber, at the time a freelance marketing specialist, was also impressed: "It was a total cliché of radio being a theater of the mind, grabbing live album tracks and presenting this concert that had never occurred. 'COZ just came along, taking all these marquee names that 'BCN had been present with during the creation of their careers, and stole the thunder."

After WCOZ's dramatic arrival, Clark Smidt began replacing the automated programming with live DJs, installing his protégé Ken Shelton in the early evening. George Taylor Morris, Lesley Palmiter, Lisa Karlin, Stephen Capen, and others were soon to follow. "Then there was Mark Parenteau," Smidt laughed. "I remember him telling me, 'I'm so broke, I've got to get hired. I had to run the tolls from Natick just to get here!'" Parenteau then suggested that Smidt hire Jerry Goodwin, one of his buddies at WABX-FM in Detroit, who had been a veteran announcer with

Mark Parenteau gets his break at Boston's Best Rock: 94 and a half, WCOZ. Photo by Dan Beach.

his "deep pipes" at several A-list radio stations around the country. Goodwin, a native New Englander, had returned to Boston to pursue a PhD in social ethics at Boston University's School of Theology. "Mark [Parenteau] seriously hunted me down. He said, 'Give up this religious zealotry you're into. Come back to where you belong!' It seemed like a good idea, so I started working for 'COZ [as] their production director and doing weekends."

WCOZ's list of album tracks and mainstream hits included songs that would become classic rock standards many years later: "The Joker" from Steve Miller, "Baba O'Riley" by The Who, Elton John's "Saturday Night's Alright for Fighting," and "Sweet Emotion" from Aerosmith. Along with a tighter playlist, 'COZ guaranteed fewer commercials, less talk from the DJs, vibrant station identifiers, and up-tempo production. Smidt called it, "Boston's Best Rock: 94 and a half, WCOZ." The mix appealed to an audience that suddenly had an alternative to the only FM rocker in town. "The first book [ratings period], fall '75, was a legendary one," Smidt recalled. "For listeners twelve years and older, wcoz jumped to a 2.9 [share] right out of the box, and 'bcn had a 1.9."

"The purists among the radio elite in Boston were horrified," Parenteau pointed out, "but the audience dug it."

"[wcoz] had a focus to their programming, and what had been an advantage at 'bcn turned into a confusion," David Bieber reasoned. "wbcn had a story to tell and their story got stolen; a *Reader's Digest* version of the station defeated them."

"When 'coz happened, it was like a pail of cold water," Tim Montgomery said. "I remember thinking at the time, 'Man to man, we *are* a little indulgent.'"

"There was so much of that self-sanctioned, self-righteous thing in radio, and at 'bcn, at the time," added Parenteau. "For instance, if it was a windy day, each jock would come in with the same genius idea; you'd hear 'Wind' by Circus Maximus, 'The Wind Cries Mary' [from Hendrix], and 'Ride the Wind' by the Youngbloods—forty-five minutes of fuckin' wind songs!"

Not only did Clark Smidt and his tighter playlist exploit the inconsistencies of a station where the individual radio shifts were self-governed, but also certain songs and genres either intentionally or inadvertently missed at wbcn became important building blocks of wcoz's programming. Ken Shelton stated, "'bcn played the Allman Brothers, but considered Skynyrd to be a cheesy bar-room rip-off. [But] people were calling 'coz: 'Play "Free Bird"! Play "Free Bird"!' 'bcn stopped playing 'Stairway to Heaven.' They must have looked at the log and realized that they'd played it five thousand times; wasn't that enough? So, they started losing listeners to the 'common man' radio station." 'bcn's jocks reacted, feebly at first, not yet grasping the seriousness of the situation. "We just started making fun of them," John Brodey remembered. "They were the fakes; we were the real deal."

"They put on the cheesy announcers like Ken," added Tim Montgomery. "Now, don't get me wrong, he's a fabulous guy and he [would have] a great career [later] at 'bcn, but back then, my God, he sounded like . . . an announcer! He had a nice voice!" But, no matter how little regard the wbcn jocks had for their streamlined challenger, the disdain was not enough to prevent wcoz's steady rise.

Aside from the listener's acceptance, wcoz also began to gain some measure of legitimacy within the music business as record labels recognized the value of promoting their releases on "94 and a half." As 'coz's ratings swelled, a seesaw battle ensued, with each station racing to obtain exclusive association with the biggest talents of the day. Most often, WBCN could rely on the relationships it had fostered over the years with record promotion people and the bands themselves, many of whom had grown up with the station. However, that was not always the case, especially when WCOZ became more aggressive in pursuing its opportunities, as when Mark Parenteau pirated The Who right out from under WBCN's nose. The original FM rocker had always enjoyed a tight relationship with the English band, dating back to even before Pete Townshend introduced *Tommy* on the air with J.J. Jackson in 1969, but that didn't stop Parenteau from recognizing an opportunity and then exploiting it to the max.

On 9 March 1976, The Who returned to play Boston Garden, and although scheduled to be on the air later that night, Parenteau still had time to see a good chunk of the band's performance before he'd have to leave. "While we were waiting for The Who to come on, all of a sudden there was this big commotion in the loge next to us; some trash was burning under the bleacher seats," Parenteau remembered.

People started moving quickly, lots of screaming, "Get out of the way! Get out of the way!" This giant column of smoke rose up and hit the ceiling of the Boston Garden, then mushroomed out and settled back on the crowd. The fire alarms went on, all the emergency doors opened, and the Boston Fire Department came in with fire hoses to get this trash fire out. People were trying to find places to sit . . . in stairways, aisles; they moved people away from the fire, but they never closed the show down. Meanwhile during all this extra time, Keith Moon, the drummer of The Who, was doing what Keith Moon always did—getting really fucked up backstage. I think on a good night, they had that all timed out so there wasn't this enormous break between bands, so he was just high enough to go on stage.

Finally the lights came down and everybody went "YEH!" The fire's was put out, we weren't gonna die, and everything was great. The Who came on and started playing, but when they went into the second song, Keith was all over the place; he raised his arms, hit the drums, raised his arms . . . and fell backwards right off the drum [platform] and out of sight! So, off they went and the houselights came back on. Everybody was all tuned up on whatever drugs they were doing and [going] "uhhh . . . what?" There was still smoke in the air, and it was cold because the doors were open; the fun had just been leached right out of it all. Then Daltrey came running [back] on and said, "It's a bummer; our drummer is ill! But, we're going to come back!" Everybody was booing, so we went backstage. I found Dick Williams and Jon Scott, the two heavies from MCA, the band's record label. They said, "Oh, Mark . . . terrible . . ."

"Listen, here's the hotline number; if you want, those guys can come over on the air and explain what just happened." So I went back to the studio and told the whole story; the phones were buzzing with people being pissed off.

Parenteau was barely into his show when the hotline rang. "The security guard called and said, 'There's a Jon Scott down here with some other people and they want to see you.'" It was Roger Daltrey, who immediately wanted to go on the air. "He explained what had happened and said, 'We're going to come back,' and gave the date [1 April]. So, the next day it was in the newspaper; the article mentioned he had come by WCOZ . . . and 'BCN had nothing to do with it. The ramifications were loud and long lasting. This was a major story, a major event at the Garden, a major band. That legitimized 'COZ as being the right station at the right time, where the action was, and a nail in the coffin of 'BCN—for what they were doing at that time. Norm Winer [must have] gone out of his mind!"

David Bieber observed that it was a most difficult period for WBCN: "The station was searching for its own meaning and what it was all about; if 1976 was the real challenge and crossroads time, then '77 was the year of twisting in the wind and not knowing where to go." Music was changing: from the radical underground rock of the late sixties and political Vietnam rants of the earlier seventies to a smoother, more mainstream, arena-rock style that sounded tailor-made for the streamlined WCOZ but was accepted with difficulty on the veteran station. Boston released its first album in September '76; Steve Miller's *Fly Like an Eagle* became one of the year's biggest hits; and *Frampton Comes Alive* spent ten weeks atop the *Billboard* chart. "A more uniform, bland tone pervaded the FM airwaves," summed up Sidney Blumenthal in the *Real Paper* in 1978. "Rock became commercial culture." David Bieber added, "You had some acts that were originally great BCN [artists] now crossing over significantly into Top 40. For example, the monstrosity that Elton John became: complete indulgence and decadence, costumes instead of content."

The only member of the full-time staff who seemed to thrive during this time was Maxanne, whose original disposition toward rock and roll had pointed the DJ toward the future way back in 1970. Her passion had helped ignite Aerosmith's career, which by 1976 had resulted in two platinum albums, a handful of hit singles, and sold-out tours. A personal and professional interest in Wellesley singer and guitarist Billy Squier had helped his band Piper get a record deal with A & M, setting up his later, and significant, solo success in the eighties. Plus, as the *Boston Phoenix* reported in a retrospective article in June 1988, "Maxanne Sartori was an original champion of the Cars. Throughout '76 and '77 she played demo tapes of the band, providing it with a ready-made audience in a key college market when it signed with Elektra." She told the *Boston Globe* in 1983, "I played the first Cars tapes so much that people at the station used to take the tapes out of the studio so I couldn't play them!"

"Maxanne was the only jock who was really rocking," remembered "Big Mattress" writer Oedipus, who was similarly excited by what he called "the nascent rock and roll scene developing in the clubs." Oedipus worked the morning show shift (paid by whatever records and tickets he could scrounge), headed home and slept all day, and then went out nightclubbing, returning to the station after last call. "I'd stay up and hang out with Eric Jackson or Jim Parry." Not only did Oedipus get to know the overnight personalities at 'BCN, but he was also there when the news department reported for duty in the early morning. "John Scagliotti knew that I got around town, so he said, 'Why don't you report on what's happening?' I was able to go to all the concerts, the cabaret shows and all kinds of cultural happenings." Oedipus began taping one-minute reports that ran in the afternoon on Maxanne's show. "It got sponsored and I actually got paid. So, for my report, I made twelve dollars a week, and I sure needed it! On Fridays I'd get to 'BCN and wait for Al Perry to sign the damn checks; that was my food." Because of his associations with Maxanne and the news department, Oedipus was not left high and dry when his mentor Charles Laquidara left WBCN for the "Peruvian" mountains in 1976. "I said, 'Charles, I should take over for you.' He looked at me and said, 'Oedipus, you'll never be hired at WBCN as a DJ!'"

With Al Perry, Norm Winer, and Laquidara gone, the rebuilding effort at WBCN began with the appointment of a new general manager, an experienced radio man named Klee Dobra, whom Mitch Hastings plucked from

KLIF in Dallas. In late January 1977, the new boss brought in Bob Shannon, another outsider weaned on radio in Texas and Arizona, to assess the station and interview to be its next program director. Shannon camped out in a hotel room and tuned in the two radio opponents to see if he could determine why a young upstart was beating up Boston's legendary FM veteran. "What I discovered after listening for awhile was that 'COZ was playing the Beatles, the Rolling Stones, and Led Zeppelin [while] 'BCN was playing Graham Parker and the Rumour. There was no question 'BCN was too hip for the room." Shannon went back to Dobra: "I said, 'With all due respect, three-quarters of the stuff I'm hearing on the radio station, I've never heard [in] my whole life. There's a lot of people that want to listen to this radio station: they tune in, [but] they hate what they get, so they tune out.'" After being hired, the new program director worked on a plan to rescue the station, concluding, "At the time everybody played exactly what they wanted, and as a result, when people were great, they were great. But when they were not in a good mood, it was awful. Plus, there was no consistency from shift to shift."

Shannon had some record cases built and then called on the staff to recommend the best albums in each category of blues, jazz, folk, R & B, and rock music. He ended up with a core library of two thousand records, which became the bulk of the station's available on-air music. Then, only at certain times in each hour were the DJs given the freedom to access the additional, and gigantic, "outer" record collection. "The idea was to reduce the library to a workable number of tunes. In the beginning they hated it, but all of a sudden the station started sounding consistent." Tommy Hadges, who had filled the morning gap following Charles Laquidara's departure, said, "We began to have an awareness and desire to look from show to show, to have some sort of flow so it didn't sound like six different radio stations." But not everyone was amenable to the changes. The first to abandon ship was Andy Beaubien, who left his midday slot to pursue a career in artist management. Original 'BCN jock Joe Rogers, who had been back at the station since 1972 operating under his new radio moniker of Mississippi Fats, also departed. "I was having more trouble with the changes than most people; I was having . . . less fun. I think ['BCN] was wonderful for a long time and I can see why it was necessary for it to change, and so it did. It isn't regrets; it was the sadness of watching . . . when all the jazz albums disappeared, all the blues albums, folk . . . . I felt, 'I'm a dinosaur here.'"

Rogers signed off to pursue a dream he had fostered—starting his own restaurant—and Mississippi's Soup and Salad Sandwich Shop opened shortly after in Kenmore Square. Virtually an art gallery that served food, the establishment featured a wall décor of huge pea pods, images of mustard, and other edibles rendered in gigantic size by students at the Museum School. "It was a lot of fun," the new restaurateur enthused. "We had real and imaginary sandwiches—peanut butter to caviar. We charged people for peanut butter sandwiches according to their height: there was a pole by the cash register, and taller people would pay more. You could get a 'Gerald Ford' sandwich: cream of mushroom soup on white bread." The pioneering disc jockey didn't forget his adventures at WBCN: "I had a sandwich named the 'Charles Laquidara,' a custom prosciutto-based Panini with provolone and tomato. But prosciutto was so expensive that we would have had to lock it in the safe at night, so we compromised with Genoa salami." He laughed, "We had a total of two customers who ever ordered the 'Gerald Ford'; but the 'Charles Laquidara' ended up being much more popular than the president!"

WBCN lost its most powerful and distinctive personality when Maxanne decided to toss in the towel, exiting after her final show on April Fool's Day 1977. "She was scared that I was going to change the radio station dramatically," Bob Shannon concluded. But Maxanne also wanted to pursue a career in the record business, which made a lot of sense since she had always tuned in to new talent. Trading in her headphones, the jock picked up a job doing regional promotion for Island Records, later working in the national offices of Elektra-Asylum and eventually as an independent promoter. Replacing such a memorable talent was an important decision, and Shannon opted to move the infamous oddball Steve Segal (now referring to himself as Steven "Clean"), who had left 'BCN for the West Coast and then returned, into the afternoon drive slot. "On the air he was either brilliant or awful—no in between," Shannon laughed. "One day he played a record by the Pousette-Dart Band and he came on the air and said, 'That reminds me of a game I used to play when I was a little kid, called "Pussydarts." What you do is take darts, throw them at the cats and try to pin them down.' The animal-rights people went nuts. I mean, they were in the front lobby in thirty minutes!"

Shannon moved John Brodey into evenings and made him music director, and then planned to switch Tommy Hadges, who had inherited

Charles Laquidara and Matt Siegel clowning around with Bill Russell.
Who's taller? Photo by Eli Sherer.

the morning shift out of necessity, into the midday slot. But, finding the right talent for mornings was a daunting task. Shannon flashed on a DJ he never met but had heard regularly on a small Tucson radio station back in '72. "Matt Siegel did a night show there, and I used to sit in the bathtub, smoke a joint, and listen to him. I laughed a lot because he was so funny. So now, I started making calls . . . but I couldn't find him; nobody knew where he was. The only thing anybody knew was that he had done the voiceover on a regional Fleetwood Mac spot for Warner Brothers." That actually was true: Matt Siegel had left Tucson and headed for LA, snagging a lucrative freelance job making commercials for record labels. When that opportunity eventually played out, he hit the streets looking for another job but couldn't find anything. With his money just about exhausted, Siegel phoned his best friend, who happened to live in Boston: "I said, 'I have to come see you,' took my last $300, and got to my buddy's house in Brighton."

Matt Siegel was now hanging out in the same city as Bob Shannon, but neither of them knew it. "There was this guy, Steven 'Clean,' who I met in LA, and he worked for WBCN," Siegel continued. "I figured I'd call him on a whim to see if there was something available because I was stone broke." Siegel phoned, and then went to the station, but couldn't find the DJ. "I was just coming up zeros. So I said, 'I'd like to talk to the program director.' I lied to them, saying, 'It's Matt Siegel from Warner Brothers in LA,' hoping the word wasn't out." But, once his unexpected visitor's name was passed along, Bob Shannon sat there stunned; the coincidence too outrageous to be real. "I told the secretary, 'Ask him if it's the Matt Siegel from KWFM in Tucson.' She came back and said, 'Yeah, that's who he is.' I couldn't believe it!"

"[Shannon] looked at me and said, 'Matt Siegel, how the hell are you?' I was amazed. 'You know me?' He said, 'I've been following your career.' I said back to him, 'I don't have a career; what are you following?'" When Shannon invited Siegel out to lunch, the DJ was floored: "I felt like I was in the Twilight Zone; my luck was finally turning."

"We were having lunch," Shannon remembered, "I was talking and watching him, and he was *really* funny. Finally I said, 'How would you like to do mornings on WBCN?' He looked at me like I was the craziest person he'd ever met in his whole life."

"[Shannon] said, 'I can give you a job today. How's $18,500?' I went, 'Uhhh, great!'" In moments, the unknown Matt Siegel became WBCN's new morning drive announcer. However, this tale of perfect timing from mid-1977 was not quite over, as Siegel related: "The last part of the story is that three weeks later, Bob Shannon was gone. Gone! You want to talk about luck?"

"I was reasonably smart [about how to fix 'BCN] when I walked into that situation, but I was in over my head emotionally," Bob Shannon confessed. "I was always the outsider, working long hours, alone—this red-headed, bearded asshole from Texas."

"Bob actually did a lot of good things for the station," Tommy Hadges admitted. "Whatever animosity existed was just a general unease of having somebody from the outside make all these changes; he wasn't considered part of the 'BCN family." When Shannon got a job offer for more money back in Dallas, he tendered his resignation. But just before climbing into his escape pod, he completed two more significant hirings, the first being Tracy Roach, a Brown University student and veteran of the campus radio station, WBRU-FM. The new jock started working weekends even before her graduation, becoming the youngest member of the WBCN air staff, and then bounced around as a full-timer before settling into early evenings.

Roach had no experience with wbcn's rich counterculture heritage, having only been thirteen years old when Joe Rogers dropped the needle on Frank Zappa back in '68. She told *Record World* magazine in 1978, "Danny Schechter did an enormous retrospective on the great and glorious past at 'bcn, which I surely have a lot of respect for. But those days are over. There are new things to be done now." As Roach settled into her job, Klee Dobra approved the deal to entice Jerry Goodwin from wcoz. Then, as the ink on Goodwin's contract dried, Bob Shannon disappeared. While his controversial six-month tenancy proved to be critical to ensuring 'bcn's future, within a few years, the young Texan's presence would be all but forgotten.

In the summer of 1977, Tommy Hadges stepped in as Shannon's replacement. Although he'd stay less than a year, he was most proud of hiring three significant employees who were destined to shape and guide the station for years to come. The new program director agreed with his predecessor that the playlist needed to be focused and consistent from shift to shift but also believed that some degree of openness had to remain. Staying in touch with the latest artists and the newest trends in music was necessary to serve this latter goal. Hadges remembered asking the staff, "What is this alternative stuff coming out?" He referred to the punk movement building on both sides of the Atlantic, with Boston being one of the point markets for the edgy style. "My roots were folk-rock, so I figured we needed to have a guy on staff who at least had an idea of what was going on, because it was definitely a different scene." As it turned out, "that guy" already worked for Hadges, writing and producing his one-minute nightlife reports for wbcn. Oedipus had been very active in the two years since he'd been brought on board by Charles Laquidara, not making a whole lot of money but diving deeply into the city's developing rock and roll culture. Oedipus explained,

> There was a different kind of attitude; songs were passionate and intense.
> There was no possibility of sitting down to this music. We didn't have time
> for all those guitar and drum solos! A scene developed with unique looks and
> black leather jackets. I'd be walking old Beacon Hill with the most intense
> purple or red hair you could imagine—just flaming! It's nothing today, but
> back then, no one did it. Plus, I had these big red Elton John–kinda glasses.
> But it was supposed to be expressive. Just like the hippies with their long
> hair in the sixties and seventies, this was another statement: you stood apart.

Tommy Hadges, WBCN's new program director, at the Prudential Studios.
Photo by Don Sanford.

I thought this music should be played [on the air], and it never would be on 'BCN. Tom Couch told me the MIT station, WTBS, accepted community volunteers to do programs, so I went down and volunteered—which I was certainly used to doing!

The radio host welcomed nearly every important up-and-coming rock and roll band to WTBS's sweltering basement studio: the Ramones, the Damned, the Jam, the Talking Heads, and dozens more. When a significant number of listeners mentioned Oedipus's show, "The Demi-Monde," in their Arbitron ratings responses, Tommy Hadges was mightily impressed; he quickly promoted the pink-haired punk into his own weekend overnight shift at WBCN. In 1998, Oedipus recounted the experience of his first show to *Virtually Alternative* magazine: "I was so nervous I stood the entire shift. I started with Willie Alexander, segued into the Clash, and played punk rock all night long. To my amazement, they didn't fire me, which is what I expected. That was going to be my statement. Instead they kept me on midnight to six."

Hadges's next mission was to convince his old friend, Charles Laquidara, to return to WBCN after nearly two years in exile. The ex-DJ's attitude typified the very rebellion that had fueled 'BCN's presence in the past, and his zaniness inspired a spirit of weirdness and humor that always made the station anything but ordinary. "I went to drag him back," Hadges chuckled. Laquidara, on a steady diet of cocaine, isolation in the woody Massachusetts suburb of Stowe, and deepening introversion, put up stiff resistance. "He was completely happy; it was quite an emotional thing to convince him that he was wasting his life away." But Hadges finally prevailed on Laquidara to, at least, meet with Klee Dobra to discuss the idea.

"We were all in a room together," Laquidara recalled, "and I was not serious about coming back. So, I just started throwing out these outrageous numbers . . . and they kept saying 'yes.' Then I went to the bathroom, did a couple sniffs of cocaine, came back and said, 'It's got to be tax-free!' They offered me a tremendous amount of money, and I said, 'My ratings weren't even that good before I left.' And Tommy said, 'Yeah, but you're a motivator; the station is just dying. We need some pizzazz, and you can do that.'" Laquidara signed back on once Hadges and Dobra agreed to his additional demands: "I got a contract they couldn't break: I asked for eight weeks [of] vacation; I said, 'No military ads.' I also said I didn't want to use my real name. I had such a great last show; I didn't want to keep coming back." So, Charles Laquidara returned to the air waves of 'BCN at the end of '77, doing a weekend show as Charles Faux Pas Bidet. "That only lasted two days; I changed it to Lowell Pinkham, a guy in my school." Still, the ideal pseudonym eluded him. Laquidara mentioned the nerdy Pinkham to a friend of his, who responded, "We had a guy like that; his name was Duane Glasscock." Charles loved it. "But I needed something in the middle. D.I.G. . . . dig? Duane Ingalls Glasscock! That was it! He'll talk with a Boston accent and be a guy that exposes all the hypocrisies around him." Laquidara lurked behind his new character much like he'd hidden himself behind the mounds of coke. Eventually, though, enough of the old self-assurance returned that he was persuaded to return to weekday mornings as himself, and D.I.G. was put on a shelf.

However, Duane would rise again. The mysterious alter ego, known as Laquidara's teenage "Six Dollar Clone" (as opposed to the "Six Million Dollar Man"), allowed the DJ to indulge his passion for acting, Duane emerging for the occasional fill-in show and a more frequent Saturday midday shift. WBCN's future music director and producer Marc Miller attested, "He would not respond to the name Charles on Saturdays, only the name Duane. That was really nuts!" Glasscock was destined to be fired several times but, like a cockroach, always popped up again, shouting his trademark "Hello, Rangoon!" to announce each return. Even though the shows were organized affairs with prerecorded bits and routines from 'BCN's inventive production staff, Duane's ad-lib style and ADHD personality, along with his (supposed) amateur ability, resulted in a true free-form pandemonium. Duane traveled to places that even his extroverted alter ego wouldn't touch; and the show always championed some specific cause or goal, like ousting Mayor Kevin White from office, giving away a dinner with the Stones (not the band, but a local couple), or taking over 'BCN to force management's hand in giving the clone more radio time.

In one memorable Duane segment, the excitable host became agitated with Arbitron, as Charlie Kendall, who would replace Tommy Hadges in the program director chair, recalled: "Duane got on the radio and said, 'We just got the ratings back and they say we have no listeners! Let's prove to them we do. I want you all to send a bag of shit to Arbitron!' I was listening and I go, 'Did he just say that?' Yes he did! And he said it again and again."

"Duane told them, 'Make sure you put it in plastic bags because you don't want the postman knowing,'" Laquidara clarified (referring to himself in third person). "He gave out the exact address of Arbitron in Beltsville, Maryland, *every fifteen minutes*."

"About two, three days later, I get a call from a friend of mine who works at Arbitron," Kendall resumed. "He said that they had gotten 14 bags of shit [in the mail] and would I happen to know anything about it. I said, 'Uh, gee, no.' He said, 'Look, you've got to stop this from happening anymore.' So, I had to tell the general manager." Laquidara recalled,

Monday came and I was doing my show. Klee Dobra popped open the studio door and said, "I'd like to see you after your show." So I went to his office and he was sitting in there with this huge cup of steaming coffee, and he was stirring it with his finger! "Uh-oh."

He said, "Charles is the consummate radio announcer; he is a professional. But Duane Glasscock is a FUCKING IDIOT! And he's fucking over! He's never to step foot in this station again! If we lose our license over that asshole, there's going to be hell to pay!"

Duane Ingalls Glasscock, the Six Dollar Clone, would eventually
die, but it didn't happen until 1989. A D.I.G. official armband of
mourning. From author's collection.

"Klee, you can't do that. Duane has huge ratings; he has a huge cult following . . ."

"Stop! You're talking about him like he's another person!"

"But you just fired *him*, and kept *me*!" Anyway, Duane was blown off the
air for three or four weeks, but by popular demand, he was back.

With Charles on board again and the bespectacled Oedipus rocking
out, Tommy Hadges's third major hire, in March 1978, was David Bieber,
the promotional maestro who had occasionally worked with WBCN as a
freelancer in the past. "'BCN never had any budget or money for marketing and advertising," Bieber said. "But Tommy and Klee convinced Mitch
Hastings that it was opportune to have a kind of resurgence." With an
actual budget in his hand, the new hire began flooding the local, mostly
print, media with WBCN's presence. He not only created ads that focused
attention on the heritage that the station had justifiably earned but also
heightened awareness of the rebuilt and recharged jock lineup. The station
now began to grapple with its radio rival in earnest. Instead of being locked
into the self-absorption that had crippled it earlier, 'BCN now understood
the politics and practices of competition and how to do "underground
radio" in the brave, new world around it.

But as Boston dug itself out from the "Blizzard of '78," and even before
Bieber had settled into his new office, Tommy Hadges arrived at a diffi-

cult decision: he would accept the offer he'd received to join WCOZ as its program director. At the same time, quite independently, 'COZ's afternoon drive personality Mark Parenteau decided he needed to move on, too. While Hadges desired the professional surroundings of Blair Broadcasting, "a more respected company," as he put it, Parenteau had grown to hate the place: "WCOZ had a corporate mindset, not at all like where I was in Detroit, which was a hippie station like 'BCN. Tight playlist . . . I had to sneak records in. You couldn't even touch the records either; they had union engineers. You were at a desk with a microphone and a phone; that was it." Parenteau met secretly with Klee Dobra at the Top of the Hub restaurant and cut a deal while Hadges worked on his own arrangements. Then, as in some Cold War exchange between East and West, the two left their respective jobs on the very same day, passing each other on their chosen paths like two captured spies swapped by their respective governments.

WBCN's newest addition had been introduced to radio at a very early age in Worcester, thanks to his mother, who was part of an afternoon women's talk show on WAAB. He'd stop by the station on the way home from school and inevitably be invited into the on-air conversation. "They'd put the big boom microphone on me, and they loved me because I would just talk. There was no stopping me; I'd just go on and on." As Parenteau got older, he hosted his own record hops on Friday nights at the Bethany Congregational Church, which catapulted him onto Worcester's premier Top 40 teen station, WORC-AM, when he was only fifteen. "I became Scotty Wainwright on the air; I was a sophomore in high school making a ton of money. I'd be on from 10:00 p.m. to 2:00 in the morning. My mother would pick me up and take me home; I'd get some sleep and be in school at 7:30. The kids that had gone to sleep listening to me were now sitting next to me in school." By eighteen, Parenteau as Wainwright, was living in Boston and taking the commuter rail to his new gig at WLLH in Lowell. When he attended the infamous Alternative Media Conference in Vermont early in 1970, connecting with some Detroit radio movers and shakers, it led to a successful stint on FM rockers WKNR and then WABX before he returned to Boston as a seasoned veteran nearly six years later. "But I always wanted to be at 'BCN . . . *the* great station."

With the departure (again) of Steven "Clean," Matt Siegel (displaced and a bit unsettled by Laquidara's return) ended up in afternoons for a short time. New arrival Mark Parenteau grabbed the evening shift, and Tracy

Roach took middays. When Klee Dobra brought in radio veteran Charlie Kendall to be operations manager (a program director with some expanded responsibilities), the team, befuddled by the defection of their boss and friend Tommy Hadges, now had an experienced programmer at the helm.

"The thing about 'BCN was that the attitude was so counterculture that it was hard to get them to focus on the fact that it was a business and we needed to be entertaining," Kendall pointed out. "I had to instill the will to win in that staff and give them a few tricks on how to do that: getting in and out of a break and how to relate to the audience in some way other than simply being self-focused and self-serving. I introduced a thing called center-staging: 'As long as every other song is by an artist that can fill up the Boston Garden, we're going to win.'" Matt Siegel didn't think the new restrictions were a big deal at all. He chuckled, "When you talk about Kendall tightening things up, it was like telling your kids, 'You really have to be in by 4:00 a.m. on school nights!'"

But not everyone agreed with Siegel. "Kendall was really a formatted radio guy; he didn't have any sympathies for what we were doing," Jim Parry charged. "It definitely tightened up a lot and became a very different station."

"Charlie Kendall didn't really like me on the air; his was more of the slick school of progressive rock radio," assessed Sam Kopper. In other words, Kendall understood and felt affinity with the WCOZ way of doing things: quick breaks, smooth DJ delivery, and tighter playlists. Kopper, who was driving his "Crab Louie" recording bus all around the East Coast to engineer as many as three live broadcasts a week for several radio stations, took the hint: "He wanted me to leave, and I was so busy anyway, we decided I would resign at the end of 1978." Kopper got a bigger bus, stuffed it with recording gear, renamed his company "Starfleet," and went on to continued success. In his defense, Kendall stated, "There were some who didn't like me, but it's what had to be done, or the place would have become a parking lot."

In May 1978, not happy with the energy level of the station, Kendall sought out a music director with a harder rock and roll edge to replace John Brodey, whose tastes ran more toward R & B, reggae, and soft rock. The DJ, operating on the air under the moniker of "John Brodey, John Brodey" in honor of the popular television sitcom "Mary Hartman, Mary Hartman," kept his air shifts but would only hang around till early '79, when he left for a job at Casablanca Records (promoting the dance music he loved but also, ironically, the primal thump of Kiss). Kendall filled the vacant slot

California governor Jerry Brown visits WBCN. (From left) Tony Berardini, Oedipus, Brown, Tracy Roach, Danny Schechter, and Lorraine Ballard. Photo by Eli Sherer.

with Tony Berardini, whom he found at San Rafael's KTIM-FM. "All Tony wanted to be was a DJ; I said, 'No, no, no, you're the music director; I need someone who rocks!" Nevertheless, he promised Berardini a radio show, and the new recruit packed up his beloved VW bus and drove east. "My first three songs on the air at 'BCN were AC/DC's "It's a Long Way to the Top," Van Halen's "Ain't Talkin' 'Bout Love," and "California Man" by Cheap Trick. I figured that was a good sampling of where I was coming from musically!" Tony Berardini would began his WBCN career with no more aspirations than enjoying his next radio show and organizing the station's music, but he'd soon be thrust into greater roles, desired or not, to eventually become one of the station's principal sculptors for a great portion of its history.

Then, Charlie Kendall took a sobering look at the news department and slashed the daily newscasts: "They were running ten, fifteen minutes long, all day long!" When he did that, "everybody got upset," Kendall recalled.

"News was reduced in size and importance," Schechter pointed out. "The consequence was that an important part of the character of the station began to be lost: that we were concerned with issues, engaged with the community, and giving airtime to people who were advocates of change. As the airtime went away, those people would go away." Kendall was not entirely unsympathetic; out of this developed WBCN's long-standing news and entertainment show called the "Boston Sunday Review." Sue Sprecher, hired as a reporter into the news department from the University of Wisconsin in 1976, liked Kendall's idea: "The station's other public affairs programs had become rather tired and almost cliché; there was a gay show, a women's show, a black show, [and] Mackie MacLeod did the prison show. I can see why Charlie, as a programmer, would take all this time and put it into Sunday morning."

"Danny told me, 'I like that. It can work,'" Kendall explained. "That got him off my back." Schechter, for his part, didn't remember being as agreeable to the idea and saw the show as a bittersweet compromise, but the "News Dissector" would be gratified to witness the "Boston Sunday Review" win the *Boston Globe* Reader's Poll as best radio talk show in the city and gain high Arbitron ratings for its time slot. A testament to its enduring strength and popularity, the "Boston Sunday Review" would continue to thrive, past the departures of Sue Sprecher and Danny Schechter in the forthcoming decade, to live on, even beyond the long-distant demise of its parent radio station.

POWER TO THE PEOPLE

As the newly remodeled WBCN, circa 1978, began to get its balance, a renewed spirit of optimism could be felt on the fiftieth floor of the Prudential as the staff dug in its heels to take on the threat of WCOZ. "It seemed like it had a conscious turnaround in attitude," David Bieber told the *Real Paper* in March '79. "Everybody was excited. We were going places again." The article went on to say that "ratings nearly doubled from spring to fall of last year and advertising revenues reached an all-time high." But another drama had been unfolding behind the scenes, cloaked in board rooms and lawyers' offices, for months. As these new developments came to fruition, not only would they divert all attention away from the battle with 'BCN's crosstown rival, but they would also threaten the very survival of the station itself. As WBCN's employees drew up battle plans and mapped out strategies, no one seriously considered that the frail-looking radio station owner in their midst was actually far more focused than they could have imagined. Challenged, yes, but still wielding all the power of his corner office, T. Mitchell Hastings redoubled his efforts to sell WBCN, and early in the year he had himself a buyer.

Thirty-four-year-old Michael Wiener and his business partner Gerry Carrus had formed Progressive Communication Corporation in 1972 to purchase the rock station KOME-FM in San Jose. Four years later, the team marshaled the capital needed to purchase another radio property, WIVY-FM, a Top 40 outlet in Jacksonville. In April 1978, as reported by the *Real Paper*, Wiener "incorporated Hemisphere for the stated purpose of buying WBCN" and "mortgaged his other two stations to raise the necessary funds." On 4 May it was publicly announced, and reported in the *Boston Globe* the next day, that Concert Network, Inc., parent company of WBCN, had agreed in principle to sell its radio station to Hemisphere Broadcasting Corporation, a subsidiary of New York–based Progressive Communication Corporation, for $3 million (Hastings craftily added another $50,000 consulting fee paid to him directly for each of the ten years following the station's sale, adding another half million to the total price). A purchase agreement was signed within a week and an application filed with the FCC to reassign WBCN's license to the new owners, who would take possession of the facility and radio signal as soon as the government approved the transfer. However, bureaucratic twists and numerable delays from both the FCC and the two companies involved dragged the approval process on for months. "As a result," David Bieber pointed out, "WBCN was in limbo. Hastings, knowing he was going to sell the station, didn't want to spend any money on it for marketing, promoting, advertising, enhancing, hiring—whatever. The [the new owners] didn't hold the license [yet]; therefore, they wouldn't spend any money either. It was a real scramble to try and make something out of nothing; we were left to our own devices and improvisational skills." Charlie Kendall's grand design to reboot the station languished, severely crippled by an utter lack of funds.

The period of limbo dragged on from May to the end of the year, and then even longer into February 1979. During that time, representatives of the United Electrical, Radio and Machine Workers of America union and stewards in the WBCN's shop aimed to secure assurances from the pending owners that their union would be recognized and the progressive nature of the station's sound and presentation maintained. The Committee for Community Access (CCA), a group of concerned citizens "interested in media and diversity in radio," according to Martin Kessel, one of its members and a former reporter for the WBCN news department, became involved, tracking the transfer process. While its endorsement was not required for FCC approval, the CCA had achieved a reputation as an organization with teeth, fighting for the media and, to that end, lodging lawsuits with the FCC to hold it accountable in serving the public interest. Michael Wiener met with CCA chairman Jack Bernstein, counsel Phil Olenick, and Kessel on 30 September for two hours at the Fenway Community Center, assuring the delegation that "his approach to radio was progressive 'in every sense of the word' and that the thing he valued most about WBCN was its heritage." The exact definition of the term "progressive" would be bandied about later, but based on this perceived sincerity, the CCA recommended that the transfer to the new owner "definitely would be in the public interest."

On the second point concerning WBCN's union shop, however, things were not so neat and tidy. Wiener continued to assert that he had no obligation to assume the responsibilities of the station's union contract when the FCC approved the transfer, so the disagreement between Hemisphere and the United Electrical, Radio and Machine Workers of America (UE) representing WBCN's rank and file simmered through most of 1978 and into the following year. Phil Mamber, field organizer of Local 262, sparred with Wiener through Hemisphere's lawyer and lodged a protest with the FCC stating that WBCN's "existing Collective Bargaining Agreement *did* require assumption by purchaser of the agreement," and recommended the government not approve the transfer until that assurance was received in writing. In the meantime, a letter to Wiener on 22 October from virtually the entire 'BCN staff stated that they looked forward to their new relationship with Hemisphere but were also committed to upholding the existing union contract. On 29 December 1978, the FCC granted assignment of 'BCN's title to the new owners, although the actual takeover of the station would not be effective for nearly two months. The decision passed the buck on the union argument to other governmental agencies, not the FCC, so when Hemisphere officially assumed control on the fiftieth floor of the Prudential Tower on 16 February 1979, that issue, the critical one as it turned out, remained unresolved.

"We were really looking forward to the new owners coming along," David Bieber recalled. "They were, by reputation, these aggressive guys from New York who owned two stations and were developing this company. They were known as people that would support the properties they acquired, and we were the biggest. They were taking the most financial chance they ever had in their career. How could they not want to invest in what they were acquiring?" With Wiener's assurances that the format would be respected and the proportions of news, public affairs, and music maintained, anticipation for Friday's takeover was high, curiosity in the staff far outweighing any foreboding over Hemisphere's stonewalling of the union. Bieber continued, "We even had this incredible, dramatic, floral bouquet with a big welcome ribbon in the lobby there to greet them as they came in to take over their new property." The arrangement wasn't even acquired as part of Bieber's miniscule promotional budget but, in fact, had been purchased with funds from an officewide collection. When Wiener arrived on the fiftieth floor of the Prudential Tower and walked in the front door at WBCN, he saw the display, thanked those present for the flowers, and then vanished quickly into his office.

Soon, the new owner of WBCN began summoning staffers, ostensibly to meet and evaluate them. Sue Sprecher, as the union shop steward, witnessed what came next. "People started getting called in and one by one were coming out saying, 'I've been fired!'"

"There was a security guard there and a lawyer," David Bieber added. "I didn't know what was coming up, so I shook [Wiener's] hand and then he basically just read from this parchment that 'we are the new owners and we have no financial obligation to you and here's a severance check—see ya later!' There was no cordiality, no sentiment; they didn't want to have any connection to you."

"We started hearing of one or two or three firings and we started freaking a little, but we expected some casualties along the way," Steve Strick, who was WBCN's acting news director at the time, told the *Real Paper*. "Then it was six and seven and eight." Soon it was Strick's own turn to face Wiener. "That whole time he never once looked at me in the eye. It was like he was reading from a Teleprompter over my shoulder. I didn't know what to say . . . I just said thank you and left."

"They took over, they fired nineteen people, [and] they violated our union contract," Danny Schechter summarized. "They were pretty arrogant and didn't give a shit: 'We bought it, we own it, and fuck you' kind of attitude. I said, 'Look, you can't do this, man. We have a contractual agreement with the company; you can't just unilaterally invalidate and dismiss it.' They said, 'Don't tell us what we can do,' and such and so forth. Then I got thrown out physically by a security guard."

"I think Danny knew what was coming up, and he may have salivated into his palm before he shook Mike [Wiener's] hand," David Bieber chuckled. Although a miniscule and symbolic blow certainly, that insult would, perhaps, be the only victory in a long and bloody day.

By the end of business hours, Wiener had terminated over half of WBCN's thirty-seven employees. This included Danny Schechter, David Bieber, Steve Strick, Jim Parry (the last remaining original jock from March '68), punk maven Oedipus, and newly arrived part-time announcer Randi Kirshbaum. Along with several of the office staffers, the new owner also canned longtime account rep Kenny Greenblatt and half the sales department, which, in a gesture of solidarity with Wiener, had ironically petitioned to leave the union in January. Once the members of sales had met with Sue Sprecher and Phil Mamber just prior to the takeover, however, they rescinded that offer, deciding that to leave the union would be destabilizing. Wiener rewarded them by slashing through their ranks. The new owner defended his actions to Dave O'Brien at the *Boston Phoenix*, who in turn reported, "His startling cutbacks were necessary because the previous owner had allowed the station to become overstaffed, and his intention was to turn a station which had 'an accumulated loss of over a half-million dollars as of December 1977' into an 'economically viable one.'"

Wiener also explained the cutbacks to Jeff McLaughlin at the *Boston Globe*, rationalizing that the firings "were based on our analysis of the station operation, including the fact that no other FM station I know of has anywhere near as many employees." McLaughlin, being a diligent journalist, checked that claim out, calling a few other Boston stations to find out if WBCN's thirty-seven-person staff was, indeed, excessive. He found that rival WCOZ employed thirty-two full-time and five part-time employees, WXKS-FM (KISS) had "somewhere between 25 and 30," and an identical amount was employed at WROR. But even if one agreed with Wiener's claim that dire economic concerns justified drastic action, the issue of how cold-bloodedly the ends were accomplished on "Black Friday" still remained. Tracy Roach told McLaughlin in the same article, "There are lay-offs in our business, of course, but Mr. Wiener had best realize that there are also procedures for negotiating lay-offs, and they start with union recognition, not massacres and union-busting rhetoric." Wiener, for his part, maintained in an earlier *Boston Globe* story on 17 February that "we know we have violated no contract. We have no contract."

Tim Montgomery, as sales manager, was never a member of the union, but at this point he'd been an eight-year veteran of WBCN since its days on Stuart Street.

> They came to me before they bought the station and said, "We want to work with you, you're our kind of guy, we think you get it," and blah, blah, blah. But the beginning of the end of my career at WBCN was that day they took over. They called people into the office, where Mitch Hastings used to be, and fired them one after another in a sort of conga line. And people were coming out, in a few cases crying, "They're firing us! We didn't know this was going to happen!" I charged in and said something like, "You don't know what you're doing; you don't understand what this station is, apparently!" Well, that was the end of me; I was supposed to be "their guy." I lasted about nine months [after that]. They didn't understand what 'BCN meant to the community or to the culture; they didn't know what they had bought.

"What they definitely wanted to do was get rid of the sales staff," David Bieber recalled. But it wouldn't be as easy as it seemed. "Basically, 90 percent of the staff was in that union, and that's the last thing Wiener and Carrus wanted." At nearly every radio or television station in America at the time, if the employees had organized themselves at all, typically there would be different unions representing different departments. WBCN's relationship with the United Electrical, Radio and Machine Workers of America, Local 262, was a rare exception. Danny Schechter pointed out, "Everybody was in the union, including the salesmen and the engineers." Out of all the departments, presumably protected together under one umbrella, the only one left virtually intact during the bloodbath was the full-time air staff. Sue Sprecher observed, "They kept anybody who had an on-air presence. [The new owners] really wanted to show the public that there wasn't any change [at the station]." As such, Charles Laquidara, Tracy Roach, Matt Siegel, Mark Parenteau, John Brodey, Jerry Goodwin, and Tony Berardini all survived the cut.

After his firing that Friday, David Bieber walked numbly to his office and then began to collect his things to remove from the station. "Charles had already done his show and was out of the building, so I called him to tell him what was going on. His attitude was, 'We've got to mobilize; we've got to do something because they can't do this,' even though he wasn't one of the ones being terminated." Bieber contacted the *Boston Globe* that afternoon, which ran a story the following morning, announcing that the entire WBCN staff, those fired as well as those who retained their jobs, would be holding an emergency meeting at 2:00 p.m. that day. A source, reported as "unidentified," told the paper, "You'd better believe that a very wide range of possibilities, including a walkout, will be discussed."

"I remember leaving that Friday and walking with Julie Natichioni from our sales department, talking and arguing about whether or not there'd be a vote for a strike," Sue Sprecher recalled. "I can remember Julie saying, 'Charles won't vote for a strike.' And I said, 'Oh yeah he will!'"

"I was influenced in a union household at a young age," Mark Parenteau said. "When push came to shove, I could have kept my gig. But it was too horrible. They would have gotten away with it if Charles and I went back. I called my dad and he said, 'Stick with the union.'"

"Nixon had his Saturday Night Massacre and he lived to regret it, and you'll live to regret this too. You don't know this station and you don't know this town," Danny Schechter hurled at Wiener as he was removed from the station premises (as reported by the *Real Paper*).

"We had been doing everything we could from the previous May or June until this February winter's day, with spit and glue and a little frosting in a can, to try and keep the station together, and they came in and just decimated it. [But] they had no idea what they were up against," David Bieber related.

"They were misadvised," concluded Oedipus, who was also a union shop steward. "They figured they'd fire half the staff and there'd be no power anymore. But they underestimated."

On Saturday, some—like Bieber, Schechter, or Laquidara—might have been fired up and ready for battle, but most of those terminated just appeared stunned, straggling into Tracy Roach's Back Bay apartment to discuss their future at WBCN, if any. Their rights under the union contract, even though not recognized by Hemisphere, were explained and discussed for hours. Wiener's hardline, confrontational, and nonnegotiating attitude seemed to indicate that the opportunity for compromise didn't exist, so the assembly eventually boiled down its options to a simple black or white choice: give up or walk out. To Danny Schechter, the catalyst in bringing a labor organization into 'BCN years earlier, it was absolutely clear what had to be done. But would other, less committed or nervous staffers agree? In the end, someone proposed a vote to strike, and the motion carried with only one dissenting opinion. The "News Dissector" was heartened: "People who were not really political got together and said, 'We have to fight this!'"

"It was an incredibly passionate time," Sue Sprecher added about the final decision. "There was not a single doubt about what we had to do."

After the vote, several in the group sat down to draft a public statement that everyone in the room could agree on. The work was finished by 5:30 p.m., and Laquidara called the radio station. The on-air jock, John Brodey, saw the hotline light flash and knew what was coming: "If the vote was for a strike, I'd get the call and be expected to walk out in solidarity." He put Laquidara on the air, who began reading the statement: "WBCN employees, including all announcers and disc jockeys, news people, engineers, creative services, office people and sales staff, are on strike . . ." Moments later, he concluded with, "We call on all WBCN listeners, advertisers, and supporters to respond to our actions. We are taking these actions to save WBCN."

"I told John," Laquidara explained, "When you leave, just be sure everybody knows that the *real* WBCN is walking out the door and that what people are going to hear after that is a bunch of pretenders."

"I said, 'Okay, I understand that,'" Brodey remembered. "Then I thought, 'Man, this is really weird; I've got a song playing and I got to walk out.' But the company had the wagons in a circle; they had somebody out there waiting to jump in."

As a managing employee of Hemisphere Broadcasting, the burden of responsibility now fell on Charlie Kendall. "The company guys came in and they said, 'No union!' I went, 'These guys are ingrained; you are not going to be able to break that union in this town.' They said, 'You just keep that station on the air!'" There was no option; if Kendall wanted to keep a job, he'd have to operate as an agent for the opposition.

"Charlie had to stay on because he was management, but he was totally supportive of the strikers," Mark Parenteau said in sympathy.

"It really killed him when the strike happened," added Jerry Goodwin. "All of a sudden we couldn't talk to him anymore. I didn't envy his position." Putting in some calls, Kendall located a few DJs who, as he put it, "were, basically, starving." The program director explained to them that they

On strike! WBCN's Mark Parenteau, Charles Laquidara, and Sue Sprecher meet the press at the picket line in front of the Pru. Photo copyright by Stu Rosner with courtesy from the *Boston Phoenix*.

would be vilified during the strike and, if the union won, then they would most likely never work in Boston radio again. Nevertheless, a few who were willing to cross the inevitable picket lines stepped forward.

Kendall himself kept the station on the air after Brodey's walkout, playing song after song until six, and then, according to the *Boston Phoenix*, "aired a taped Queen concert, followed by a taped Blondie concert (an affair at the Paradise hosted, ironically, by Oedipus), followed by a taped Steve Miller concert. It wasn't until after midnight that the actual live voices of strikebreaking announcers began to come over the airwaves."

"There was no automation, so it had to be live jocks," Kendall mentioned. "I remember one guy came down to the station who had one arm!" Nevertheless, the applicant demonstrated that he could cue up a record and perform the other required tasks to run a radio show, so Kendall impressed him into his ragtag group of replacements. Michael Wiener arranged for a few pros to jet in from his other properties in San Jose and Jacksonville, and soon the alien voices had taken over the high ground. That they were thrown off somewhat by the sudden reassignment could be heard from time to time: Alan MacRobert in the *Real Paper* reported that one of the announcers forgot where he was and identified the station as "WBCN, Chicago."

While Michael Wiener sat in the nearly empty WBCN office with a skeleton crew of Kendall, Tim Montgomery, the chief engineer, a worker in the accounting department, and the motley assortment of replacement DJS, the striking staffers got right down to business, organizing themselves with the help of Phil Mamber from the UE. A list of demands was drafted, including recognition of the union and reinstatement of all fired employees, and then presented to Hemisphere's president. Wiener summarily rejected its arguments, telling the *Boston Globe*, "The union was recognized by the previous owner, but that contract is non-assignable." Matt Siegel acknowledged this in a prepared statement, saying that the FCC informed their union that the new company "need not accept a contract negotiated by the old WBCN, but it must recognize the United Electrical Workers, the union that has been certified by the federal government as our legal representative. This, the new management has refused to do. Rather than sit down with us to resolve our differences, they choose to misrepresent our cause and malign our union."

By Monday, 19 February, at 9:00 a.m., the entire group of striking employees, with a cloud of supporting station volunteers and concerned listeners, had thrown up a picket line in front of the Prudential Tower and then held a press conference to announce the strike in greater detail. The following morning, both of the city's premier dailies, the *Boston Globe* and the *Boston Herald American*, featured articles about the action, including nearly identical photographs showing the circle of picketers braving the February chill, with producer Marc Gordon in the foreground holding aloft a sign that read, "Don't Destroy WBCN." Both of Boston's foremost weeklies, the *Real Paper* and the *Boston Phoenix* would soon chime in with their own in-depth articles, thus beginning a nonstop rush of media attention, including scrutiny from national stalwarts the *New York Times* and *Rolling Stone* magazine. "There was such participation by both local and national media," David Bieber recalled. "People like Jeff McLaughlin of the *Globe* covered it almost on a daily basis, which was unprecedented."

While Phil Mamber and the UE filed an unfair labor practices suit against Hemisphere with the National Labor Relations Board, the rank and file of Local 262 strategized. Danny Schechter realized that it was a long shot, but in this situation, their union actually had a chance to win. Gathered under one collective bargaining agreement, the WBCN staff, with all its departments, displayed a unique strength that made victory possible. Schechter explained, "It's completely unusual in broadcasting because of the craft union structure, [for instance,] the engineers in one union and the announcers in another. There's never really the unity [needed] to beat the company; they could play different people off one another." With everyone gathered together in the UE, the WBCN strike committee began to set up responsibilities based on its many members' unique abilities. Jerry Goodwin recalled with a laugh, "I was on the 'knee-busters' committee. Our job was to go to the Sheraton next door to 'BCN, where all [the replacement DJs] were staying, and wait for them to come out of their door and suggest that they just get in their car and hit the road, as simple as that. We threatened to ruin their car . . . but never did. We didn't have to; they were outta there!"

"It was called the 'Dirty Works' committee," Oedipus clarified. "We would call those guys up on the hotline all the time, harass them, and a lot of them walked." The punk jock also performed another, more surreptitious and clandestine job as an actual WBCN secret agent. "I would meet with Charlie Kendall to find out what was going on over on the other side. We would meet downtown in a movie theater and sit next to each other. It

Oedipus lists the strikers' demands while a concrete overhang protects the crowd from a cold February rain. Photo by Sam Kopper.

really was like 'Deep Throat [from the Watergate scandal].' I saw one of my favorite films, *The Warriors*, three times with Kendall."

The other committees were more high profile and, somewhat, less scandalous. David Bieber fed the media a steady diet of information from the strikers and organized support from his contacts, while Tony Berardini, as music director, worked through his Rolodex. "It was my responsibility to talk to the guys from the record labels and see if they would support the strike, and they all did!" Berardini became the station's mouthpiece to all the national music trade magazines, including *Radio and Records*, *Album Network*, the *Gavin Report*, and *Record World*. He also left Boston midstrike and traveled with Kenny Greenblatt to the *Radio and Records* convention in Los Angeles, where he was given generous podium time "in a room with seven hundred people" from the music industry to explain and urge support for his union's actions.

The striking sales representatives received the task of cutting off the flow of money to the station. Sue Sprecher explained, "I think they were the ones put in the toughest spot of all. They had to call up their clients and ask them *not* to advertise on WBCN, after years of working on those [relationships]."

"One of the key factors was that we reached out to the advertisers and told them that this was not the real 'BCN," David Bieber pointed out. "If they were spending their money, they weren't getting their money's worth." In mere days, the sales staff achieved remarkable results; by Wednesday the 21st, Boston University's *Daily Free Press* reported that most of the advertisers on WBCN had thrown in their support for the strikers and withdrawn their commercials. Jim Parry told the student newspaper, "Most advertisers withdrew their spots because the station they bought isn't what they're getting." By the end of the strike's second week, the only major advertiser yet to pull commercials off the air was Don Law. Since Charles Laquidara had known the concert promoter for many years, his assignment was to convince the impresario to follow suit. "That was a big deal," Steve Strick said. "Charles really worked on Don." To be fair, the *Real Paper* revealed that Law provided the Orpheum Theater at cost for a strike benefit concert on 4 March [and another a week later] and then reported, "But he has yet to match the extraordinary steps taken last week by other local corporations such as Brands Mart, Rich's Car Tunes, Tweeter Etc., and Strawberries. In addition to removing all advertising from the struck station, these companies have also donated *paid commercial time* on other radio stations to advertise the striker's benefit." As published criticism began to mount, Law fell in line after realizing how important his company's action would be to the strikers' efforts: "I spent a lot of time with Charles and I was responsive to what he said. We followed his lead and pulled our commercials."

"It was one thing for the newspapers to give [the strike] coverage, which they did, but they [also] ran ads in our support at no charge," David Bieber marveled. "I was connecting with people like Peter Wolf and the J. Geils Band to run ads in support of the strikers and the listeners who marched in the picket lines, and to do concert benefits for the people who were out of work. Don't forget, this was February 1979, dead of winter, freezing cold." Within days of the walkout, the J. Geils Band had published "An Open Letter" to Michael Wiener regarding "The Strike to Save WBCN-FM," insisting "that any station endorsements made by members of our band be immediately removed from the airwaves until negotiations are completed." Peter Wolf concluded the letter by writing, "Personally, as a former WBCN disc jockey, and now as a listener, it saddens me that such estrangement between management and staff has occurred at a time when I feel the station was sounding better than ever." Then there was the very quotable closer: "P.S. The only scabs I dig are the ones on my elbow." In short order, all of Boston's biggest bands had published their own manifestos of support, including Aerosmith: "These people need us and we need them"; the Cars: "We will never forget WBCN's contribution to us. Let the Good Times Roll . . . Again"; and Boston: "There's no place like home and WBCN is a part of our home. Let's keep it that way." A consortium of local band managers, club owners, and musicians joined forces to declare their support in the press, and then backed up that statement by offering their talents and services to the strike committee.

Those offers were swiftly organized since there was an immediate need for funds to buy groceries for striking employees, pay their bills, and avert the pressure of being forced to consider work elsewhere. The most visible benefit concert was the Orpheum Theater show on 4 March, featuring James Montgomery, the Fools, the Stompers, and Sass. All the bands worked for free, and even the lighting and sound technicians, with all their equipment, refused a paycheck. Members of the striking 'BCN staff took the stage between acts to plead their case and thank the listeners who had stampeded down to Tremont Street to sell out the old theater. Then the J. Geils Band capped the evening with a high-octane surprise performance. This Sunday-night show was so successful that the strikers planned a second event for 11 March, featuring other A-level Boston bands: Robin Lane and the Chartbusters, Private Lightning, and Pousette-Dart Band. Using his relationship with the Rat in Kenmore Square and contacts among the Boston punk community, Oedipus spearheaded a weeklong series of nightly benefits featuring the cream of the new "underground" scene, while a major North Shore fundraiser at the Main Act in Lynn featured the Neighborhoods, Human Sexual Response, and more. FM, the movie depicting a fictional radio station's labor struggle, was given special benefit showings at the Orson Welles Cinema in Cambridge. "The movie depicted a strike at a radio station purported to have the same kind of spirit WBCN had," David Bieber observed. "But, in ninety-minutes, how do you tell a tale comparable to what we went through . . . a real, ferocious, on the street battle. No way FM could do justice to what we endured in the end." But how appropriate it was that a Hollywood yarn about a West Coast radio strike would help to fund a similar, but very real, drama unfolding in Boston.

With the staff locked out, and throwing up daily picket lines in the broad plaza in front of the Prudential Tower, the nerve center of the newly formed

ON STRIKE!
A BENEFIT CONCERT
TO SAVE WBCN!

pousette-dart band

Robin Lane and the Chartbusters

ORPHEUM THEATER
SUNDAY, MARCH 11 7:30 P.M. $7.50

- TICKETS AVAILABLE AT THE BOX OFFICE, ALL TICKETRON OUTLETS AND ALL STRAWBERRIES RECORD STORES.
- PRESENTED WITH THE COOPERATION OF DON LAW.
- SOUND PROVIDED BY TIMBER AUDIO.
- LIGHTS PROVIDED BY VIRGO LIGHTING.
- SPECIAL THANKS TO BRANDS MART, GOOD VIBRATIONS, RICH'S CAR-TUNES, STRAWBERRIES, TECH HIFI AND TWEETER ETC. FOR PROMOTIONAL ASSISTANCE.
- FOR MORE DETAILS, CALL THE WBCN LISTENER ALLIANCE, 739-2121.

The second benefit concert at the Orpheum Theater. Boston's bands help keep the striking employees afloat and the fight alive. Courtesy of Clint Gilbert.

WBCN Listener Alliance was established in a home with donated space on Braemore Road in Brookline. Clint Gilbert, a part-time manager of the expatriated Listener Line, designated a recent recruit named Paul Sferruzza (later known to 'BCN listeners as "Tank"), as the person on location who was in charge of the volunteers. "Darrell Martinie's agent was the guy that came up with that venue," Tank recalled. "We had our little work area, a bathroom, two bedrooms, and a kitchen space. People who had worked the Listener Line [in the Prudential] came in and worked their shift in Brookline." Gilbert mentioned, "We got three or four phone lines in, and once we were able to get the new phone number out, 617-SEX-2121, they were ringing!" David Bieber, working with his sympathizers in the print media, managed to obtain nearly daily ads in the papers listing the WBCN strike events and the new phone number to replace the so-familiar 617-536-8000, although the more discrete numeric ID, 617-739-2121 (without the SEX part), was the one that made it to print. Tank pointed out, "The whole focus in Brookline was on the strike and raising money so that guys like Matthew Wong, who was a veteran sales person, could pay his bills." Clint Gilbert shook his head at the memory: "The listeners were so willing to help; the caliber of what people were offering was nothing like I've ever experienced. 'Just tell me, what do you guys need?'"

As donations poured into Brookline, and some listeners just pressed cash into the hands of picketing volunteers at the Pru, another drive was on to demonstrate to the new owners just how much the community supported 'BCN. Listeners were urged to send Michael Wiener and Hemisphere Broadcasting letters of protest, or flood its Prudential Tower office with calls. Dozens and dozens of station fans converged on Boylston Street daily to augment the circle of striking protesters and march in solidarity. A "Duane Glasscock Victory Motorcade to Save WBCN" was organized for Saturday, 24 February: "A magical mystery tour of Boston," according to the leaflets placarded all over the city by droves of volunteers. The *Boston Sunday Globe* reported the next day that eighty cars, "many festooned with posters, balloons and banners," accompanied by a police escort, staged a peaceful, promotional drive through the city during a hard rainstorm. "My car was 'The Big Mattress' car," Tom Couch recalled with a laugh. "I had a Honda Accord, and we tied a big, old mattress on top of it."

Halfway into the second week of the WBCN strike, Mitch Hastings stepped off a plane at Logan Airport, after having vacationed in Bermuda

Motorcade promotional flyer. Go for it! Courtesy
of Tony Berardini.

following the station's sale. The *Boston Phoenix* caught up with the former
owner and asked him what he thought of the furor that had erupted in his
absence. Hastings, who appeared fully aware of the strike, was nevertheless shocked at what had transpired. "It's nothing like what we expected.
We thought we were selling the station to a company that would carry
on our programming and staff with a minimum of changes. I talked to
him [Wiener] today and I get the impression that he still wants to carry
on WBCN's image, and standing, but he doesn't quite know how to do it."
A softening attitude from the top of the Pru? It certainly seemed so. The
article revealed, "Faced with the continuing loss of 'BCN advertisers and

complaints from long-time listeners about the absence from the airwaves of their favorite personalities, Wiener did agree to a sit-down session with representatives of Local 262 of the United Electrical, Radio and Machine Workers of America." Jeff McLaughlin at the *Globe* wrote, "The union, led by field organizer Phil Mamber, refused to back down from any of its demands: Reinstatement of every staff member, recognition of the union for collective bargaining and negotiation with the union over any layoffs." Wiener, in the article, admitted that the loss of advertising revenue was "beginning to really hurt."

Then, in a moment of perfect timing, as round-the-clock bargaining sessions stretched on for a week, Arbitron published its radio rankings for January 1979. The figures showed that immediately prior to the strike, WBCN's popularity with all Boston-area listeners older than age 12 had rebounded into the number 6 position in Boston with a 4.7 share. At the same time, WCOZ's ratings had softened to a 4.3, earning that station the number 7 slot. This news prompted Wiener to redouble his efforts to settle affairs at the negotiating table and move on. In swift order, the two sides finally hammered out an agreement, and the union dropped its suit in response. When the sudden announcement emerged that workers would be expected to report back to work on Monday, 12 March, everyone suddenly realized, with a start, that the WBCN strike was actually over.

Jubilation exploded as the amazing news spread. "I never expected to win," Steve Strick admitted. "I totally expected that all that protesting was just for show, and that at the end of the day we'd basically be looking for jobs. The fact that they hired us back was huge . . . history making!"

"It was truly one of the highs of my life, the night after we won the strike," Sue Sprecher said. "I remember being at Mark Parenteau's place and we were all singing 'We Are the Champions!'"

Danny Schechter, channeling some (very appropriate) hippie idealism that harkened all the way back to the scribing of America's founding fathers, wrote in the strike victory press release, "We hope that our efforts can be an example to others in the Communications Industry—to show everyone that unity is strength; that colleagues should care about each other and support each other; that principles are worth defending; and that working people do have rights which are worth struggling to secure." The already-scheduled 11 March solidarity concert at the Orpheum Theater was hastily, and happily, relabeled a "Victory Benefit" as a humbled, and

Victory onstage at the Orpheum Theater, 11 March 1979. WBCN's finest hour.
Photo by Eli Sherer.

candid, Michael Wiener told the *Boston Globe*, "I'm thrilled the strike is over. I've never seen anything like it. Clearly, I miscalculated—to mount an effort like they did is just extraordinary. What can I say? I'm happy as hell we're all on the same side now."

The *New York Times* reported the resolution of the strike on 13 March, with Michael Knight writing, "It was a famous victory, everyone agreed, although no one could agree on exactly what had been won aside from the temporary reinstatement of 19 employees who had been abruptly dismissed when the station, WBCN, changed hands last month." But much *had* been won, most notably the reality that the radio station had beaten back a hostile takeover, remapping the battlefield into a cautious cooperation between two suspicious parties. The armistice validated that a business acumen found on the corporate level should and could work together with a passion found on the street level to serve the public interest. The *Times* article went on: "Michael Harrison, the publisher of *Goodphone Weekly*, an industry newsletter, called the strike settlement a milestone for those who believe a radio station is 'part of a community and not just a collection of wires and turntables, or a business like any other.'" Looking toward a new decade, there were many great things in store for what would become known as "The Rock of Boston." But as lofty, acclaimed, and noble as some of those future achievements would turn out to be, WBCN would never be as unified and purposeful as in those twenty-one days in the winter of '79. It was, and would always remain, the station's finest hour.

Fly low, beat the radar. May the wind at your back not be your own.
And remember, wherever you go, there you are, so take us along.
TONY BERARDINI, RADIO SHOW CLOSER

"THE REALLLLLL WBCN, BOSTON!"

Michael Wiener and Gerald Carrus sat in their office on the fiftieth floor of the Prudential, looking down at the twinkling skyline of the city that had rejected them. As the two owners sat stunned in their defeat, even while publicly claiming a victory, they had to admire the fortitude of their former opponents, whom they were now inheriting, like it or not. As the two pondered their next step, the ex-strikers below headed off, arm in arm, from party to party, chattering excitedly in their groups, to continue the revelry long into (or through) the night. But Monday would soon arrive, and other than the fact that they had a job to report to, nobody knew exactly what to expect once they got there. Although the outright sacking of half the staff had been forestalled, both sides realized that a reckoning still waited. Plus, outside of WBCN's bubble, the challenges that had threatened the station before the strike remained; WCOZ, the greatest radio threat 'BCN would ever face, was still very much an immediate danger. And soon, that relentless competitor would become far more dangerous as it converted, Transformer-like, into something larger and wilder: a radio beast known as "Kick-Ass Rock and Roll."

Producer Steve Lushbaugh created a famous radio station sounder that blasted out, in Godlike magnitude, "The Reallllll wbcn!" across the air every hour or so. Like a marquee and spotlights proclaiming the presence of a major event inside a theater, the blaring announcement trumpeted 'bcn's official (and authentic) return. The employees, flushed with success, reported back to work on Monday, 12 March, finding that the environment at the station was far from unpleasant. Charles Laquidara's first poststrike "Big Mattress" became a nonstop party as listeners phoned in their good wishes, friends and supporters were thanked on the air, and Laquidara established the Hot Chocolate hit, "Every 1's a Winner," as a victory song. "There was a harmonious reentry," David Bieber remembered. "There was a big party at a roller rink, catered top shelf, where everybody got skates. Mike Wiener and his wife Zena were there, Gerry Carrus . . . there was really a good feeling. It was like that refreshing breeze that happens after a hurricane blows away all the bad air."

However, although everyone smiled and acted cordially, there remained serious differences with the new owners, who still maintained that the size of the station's workforce needed to be whittled down. The *New York Times* reported, "Michael Wiener praised the settlement as a victory, although it represented a repudiation of his policies and a delay, at least temporarily, in his attempts to cut costs at the station." Even the union's official "Strike Victory" press release, largely penned by Danny Schechter, acknowledged the latter point: "The union recognizes our need to review our staff requirements. And we are determined that any necessary staff reductions will not be effected in a capricious manner."

"We had a strike settlement that we signed to end the walkout . . . which just meant that we went back to work under the terms of the previous contract," Tony Berardini explained. "Then we had to sit down and work out a new contract . . . which was really something. As you can imagine, everyone was all fired up. At that first contract meeting with Phil Mamber and the new owners, about thirty people—jocks and sales and news—all showed up. It turned into a three-ring circus; the new owners just got pummeled! Mike and Gerry had the most pained looks on their faces. It was brutal; [there were] several hours of recriminations and venting, and

The folks on the other side of 536-8000, the WBCN Listener Line
(circa 1980). Photo by Eli Sherer.

all they wanted to do was move forward and get the station back on track.
But it eventually smoothed out."

Against the backdrop of the station's familiar voices triumphantly reas-
suming their positions on the air, the inevitable trimming began. The first
station casualty after the strike was Clint Gilbert, the employee responsible
for staffing and running the Listener Line, who was let go by Wiener, ostensibly for his activities with the Committee for Community Access (CCA).
As expected, John Brodey, who had been on the docket to leave even before
the union action, soon put his plans in motion and exited for his job in the
record industry. Jim Parry, the one DJ left at the station who could trace
his radio genealogy back to the first nights of WBCN's "American Revolution," unexpectedly drew a pink slip. "I was not pleased with the end," he
explained sourly. "They fired a couple of people in each department: me
in programming, Matty [Matthew Wong] in sales, someone in the office;
it was, like, four people. But I was getting kind of [fed up] anyway, it was

really changing . . . a lot. The guys that came in knew how to run radio stations, and it became very different." Charlie Kendall soon caught an express elevator of his own off the fiftieth floor despite working for the company and achieving a great deal of success before the strike. "When I first took the job, I said, 'Look, I'll work for x amount of dollars, but you've got to give me 3 percent of the profits.'" [After the strike] Wiener and Carrus called me into the office and said, 'Look, we'd like you to stay here, but no percentage points. We don't want any partners.'"

"I didn't know about that deal, but the reason Charlie had to leave anyway was that nobody could trust him," Tony Berardini rationalized. Although the smooth-talking radio pro had helped the union during the strike, he had, after all, been Hemisphere's primary asset to keep WBCN on the air and functioning during the battle. "There was always going to be that division between him and everyone else. A program director needs to be a leader and you can't lead if people won't follow you."

With Charlie Kendall out the door, Wiener and Carrus needed a replacement, so they approached Berardini. "I said no. At that time, the image of being part of management two months after the strike and aligning myself with the people who had caused it was just anathema to me. There was no friggin' way. I just wanted to be music director and do my shift." Word about the owners' offer got around the office, though. "One day, I was in the music room; Charles and Mark came walking in. They locked the door! 'Tony, you *have* to take the program director job.' I said that after that long strike, I was back to doing what I liked doing. They said, 'No, you don't understand. If you don't take that job, they're going to bring in some guy from the company, some asshole from New York, who's going to want to change everything; he's going to want to tighten things up." Berardini smiled at the memory. "What was I going to say to two on-air leaders of the radio station who were basically telling me that they'd decided that I was doing this? I told [Wiener and Carrus] that I didn't want to wear a tie and that I wanted to keep an air shift, because [typically] the PD at WBCN didn't do a regular weeknight show. Gerry Carrus looked at me like I had two heads and said, 'That's fine; you can do an air shift. And don't wear a tie; I don't care.'"

During Berardini's first week in his new office, he'd hardly figured out where to put the desk lamp before being pulled into a disagreement between Charles Laquidara and Matt Siegel about an incident on the air. "I can't remember if it was over Mishegas [a game show segment] or something, but they were having a fight." The quiz show segment at the end of every "Big Mattress" had grown to mammoth proportions, filling out the last fifteen minutes of Laquidara's shift and serving as the segue into Siegel, who'd inherited the midday show. "It took a good hour to deal with this since Matt had to keep running in and out of the studio because he was on the air." While Tony concentrated on mediating the argument, a call came in from the receptionist. "She said, 'There's a guy who has an appointment waiting in the lobby. His name is Howard Stern.' I had no idea Charlie [Kendall] had made this appointment, and I certainly didn't know who the hell Howard Stern was. I said, 'Please apologize to him; tell him I'm dealing with an on-air issue and have him find a seat.' Then he waited a whole hour." Eventually, Berardini was able to show Stern into his office. "Of course, he wasn't the big star yet; he was just looking for a gig at the station. But what was I going to do? I already had Charles Laquidara in the morning, Matt Siegel in middays, Mark Parenteau in the afternoons, and Tracy Roach at night. This was after the strike, and we had one of our best [ratings] books ever. I had no place for this guy. What was I going to give him, weekends?" Of course, Stern would eventually return as a conqueror at the head of his own massive radio empire, but regarding this initial episode with the fledgling "shock jock," Berardini just had to smile and admit, "Who knew?"

Near the beginning of June 1979, Tony Berardini needed a part-timer to run the station from the twelve strokes of midnight to dawn on Sunday mornings, and hired his first new employee. He found me at the same place Joe Rogers, Tommy Hadges, and Oedipus had all trained: MIT's WTBS-FM (which had become WMBR only one week earlier). Berardini made his selection, I suppose, because I knew a bit about music and could speak fluent AC/DC. On the air that first night, I was speechless for six songs, but after I pulled it together, my new boss walked in and threw me the two beers left in a six pack that he and Tracy Roach had been sharing in his office while listening. I suppose it was his way of saying I passed the audition.

The radio station got back in gear as the daytime jocks resumed their shifts, and Berardini promoted Oedipus into the nightly 10:00 p.m. to 2:00 a.m. show. With the ousting of Jim Parry, Jerry Goodwin settled into the workweek overnights, unveiling a radio character who would become the offbeat nocturnal companion of many a fellow traveler on the third shift. "The 'Duke of Madness' came on right after the strike," Goodwin

Jerry Goodwin, a rock and roll president? "Bonus Bucks": a 1981 sales department credit incentive program for advertisers. Courtesy of the David Bieber Archives.

explained. "I had played Fireside Theater in Detroit and they did this faux spot for 'Duke of Madness Motors,' so I loved that name and thought to myself that I'd use it sometime in my life. Sure enough, the occasion came up." But Goodwin, as the "Duke," didn't just quietly ease onto the airwaves at 2:00 a.m.; he launched an audio assault from them. The improvised introduction to the show, woven every night from a near-cacophonous mix of music, sound effects, voices, and "drop-ins" from movies or television, could go on for as long as twenty minutes. "Wasn't that crazy? I'd make it up as I went along, and it all depended on how inspired I was."

Goodwin loved the overnight hours: "There was something magical about doing a show out of that studio and being up in the fiftieth floor at night when that board would light up with all the different colored buttons on it. The lights were low, the 'fatty' was burning, and you could look out past the Skywalk down South Shore to Quincy. You were the king of the world, seriously, at the top of the Hub . . . the top of the universe." Goodwin renamed all of his nocturnal coworkers, christening a cast of characters that inhabited the dark shadows of his shift, including Listener Line volunteer Jutes Leedon, who would become his lifelong companion and future wife but was known to radio fans as the Contessa Irina DeMarco. She was joined on the line by "Judy in Disguise" and Goodwin's first producer, Richard Morris, who became Doctor Thorazine. "I had a guy answering phones that I called Captain Backwards, a funny, funny, guy who could actually speak backwards, fluently. I loved him for that. He'd talk backwards to all these people and do promos for me . . . everyone thought he was some foreigner, for God's sake. Yeah, the overnight people were the best because they were crazier than hell . . . and that's why they were doing overnights!"

Long after the parade of gray suits had abandoned the Prudential Tower to commute home, the security detail found a very different type of client seeking to wander back into the premises, calling up on the hotline to announce them and wait for Goodwin's approval. "The clubs closed at two, so that meant all the bands and people who were driving home would say, 'Let's go by and see the "Duke!"' They just loved coming to the station and laying down on the floor or sitting up against the wall in the [record library], just chillin' and listening to the stuff I was playing. They'd come to help me out, if you get my meaning . . . they'd lay out some stuff that was just nuts!" Occasionally Goodwin would host a personality on the air. "Once, Parenteau brought in Mitch Ryder . . . you know, Mitch Ryder and the Detroit Wheels? Mitch was a surly son of a bitch and Parenteau thought that if he brought him to the station while I was on the air, we would reminisce about Detroit radio. [But] Mitch [just] sat there. It kind of went, 'Mitch, how you doin'?'

'All right.'

'Yeah . . . good. You still living in Detroit?'

'Yeah . . .'

'Well, thanks a lot, man.'" Goodwin threw up his hands in frustration. "He wouldn't talk! But, Parenteau was brilliant when he got an artist who was reticent, or not forthcoming. He had one question that I loved and I would wait for him to ask it: 'Well, okay, let's get down to it. What happened to that last album?' That was his question, and whoever it was would just, all of a sudden, spill their guts: 'We couldn't get rehearsal time' . . . blah, blah, blah and off they'd go. That was the magic question and I would wait for it. Just brilliant, man."

Since most self-respecting rock stars slept late, Mark Parenteau's 2:00 to 6:00 p.m. shift was the magic one for interviews, with dozens of the most famous artists (and con artists) walking through the studio. Sue Sprecher recalled, "My favorite 'BCN memory was working at the top of the Pru, and there was a knock on the [back] door. I opened it up and it was John Belushi! He goes, 'Can I go on the radio?' I said, 'Let me check, but I think, probably yes!' Parenteau was on the air, and I walked in and told him, 'John Belushi is outside and wants to go on the air.' He didn't believe me, but then he looked . . ."

"You could see right from the control board, through the Listener Line window, at who was coming in the back door," Parenteau said.

There's Belushi, and a couple of local cokeheads. He spent a lot of time on Martha's Vineyard and knew the station quite well, was in Boston for the day, and hooked up with all the wrong people. They were wired to the tits and decided to come by unannounced and unexpected to 'BCN to do an interview. Belushi had become a superstar with *Saturday Night Live*, and [in 1978] *Animal House* came out and he was huger than ever. So, the door opens, and within five seconds I snapped on the microphone, just overreacting: "John Belushi, I don't believe it!" It was like Christ had just walked in the door because he was so big and everybody wanted a piece of him at that time.

During commercials, Parenteau escorted Belushi to the back of the record library, where "treats" awaited him. "It was a memorable day, and I don't even think we have any pictures because it was so unexpected."

"I'm not Mark Parenteau; I don't know how to do interviews," Charles Laquidara confessed. "So Frank Zappa was sitting there, on the air, and I was trying to casually make conversation: 'Why are you wearing these red tights?' He said, 'To hold up my balls.'"

The Clash and Oedipus got together on WBCN after the band's show at the Orpheum in September 1979. They hit it off fabulously. By the end of the interview, the members of England's foremost punk group were singing along uproariously on the air with the Village People's "Y.M.C.A."

"Jesse Colin Young came in to play on the air and he was famous as a real 'we're all brothers, we're all water, love, love' kind of guy," Laquidara recalled. "Tom Couch ran into the studio to give me a tape of something I'd asked for, and he tripped over Jesse's guitar and knocked it over. Jesse called him a fucking asshole and told him to be careful where he was going. I said, 'You know, I don't really want to talk with you anymore,' and I asked him to leave."

During an interview with Oedipus, Jean Jacques Burnel from the Stranglers baited the listeners with statements like "The only thing Boston is good for is scrod!" The flood of ensuing on-air phone calls were rife with retaliatory insults and expletives, an alarming number of f-bombs from both sides prompting the DJ to close down the lines. Burnel taunted him haughtily: "You're just afraid that your big building will get blown up!"

The Cars share driving tips with Mark Parenteau. Photo by Eli Sherer.

Oedipus replied with an uneasy laugh, then, "Well, if it hasn't been blown up already . . ."

"Hunter S. Thompson came in and he was really pissed off," Charles Laquidara observed. "He had a new book out and his publicist had set up this morning interview. 'I've been up all night; let's get this fucking thing over with!' So I said, 'Hunter? Your, uh . . . coffee is in the other room; go in there with Tom Couch.' So he disappears . . . for an hour! He's doing, uh . . . Columbian coffee, shall we say, in the production studio, making all these promos for us that started with, 'Hi, I'm Hunter S. Thompson and WBCN is the best station I've ever heard in my life . . . but I can't tell you why!'"

Poststrike, Matt Siegel, his brief spat with Laquidara long forgotten, relished each day's opportunity to have fun with the morning jock during their "Mishegas" crossover. "I was the dry one, Charles was the straight guy; we would riff back and forth and I would just take shots at him. I learned a lot about styling back then, taking that pause before you told the punch line. That was the beginning of me actually finding my radio voice; Charles was the one who gave me confidence." Siegel continued with a smile,

If you mention, even in passing, that somebody needs something, [then] he's on the phone just wheelin' and dealin'; but, by the same token, [Charles] never bought *anything* when he was [at 'BCN]. He lived his whole life on this sort of barter system and everything was free at his house. [For the game show segment "Mishegas"] he would book a contestant . . . a category of people, and it was always something for the house! "Hey we're going to play with plumbers today, so call me if you're a plumber." He'd get a plumber on the phone, and they'd work out some sort of deal . . . then the guy would fix a toilet and get tickets to a Don Law concert!

Siegel was hilarious enough on his own show, working the comedy and pulling pranks as he gained confidence. In September '79, WBCN obtained the exclusive early release of Led Zeppelin's first new studio album in three years, and anticipation ran at a fever pitch. Any disc jockey would have been thrilled just to have the opportunity to debut such a long-awaited monster, but Siegel went for the fake instead. When the big moment arrived to play *In Through the Out Door* for the very first time, he substituted a generic disco number, allowing the sinuous rhythms and female "Oooh-Ooohs" to pulse just long enough to panic even the most casual Zeppelin fan. Then, finally turning the microphone on, he wailed, "I can't believe it! That's the last straw—I'm quitting! Led Zeppelin has gone disco! I've had it with rock and roll!" It seemed to be a goof, but for a long uncertain moment, thousands of Zeppelin fans sat paralyzed by their radios, not quite sure. They wouldn't be completely reassured until a minute later when the joker started up the real thing on turntable 2. Talk about timing.

Then, there were the famous Lunch Songs, a tradition begun on Matt Siegel's "Mighty Lunch Hour" at noon, even before the strike. Tom Couch gave credit to Steve Lushbaugh as the studio genius who did the first song parodies on the station. "He was a revelation, had a million voices. I was, like, 'How did you get rid of the words in the original song [so you can substitute your own]?' He messed around with the plugs in the back [of the control board] and I learned from him." Soon, noontime spoofs with titles such as "Lunch-itis" (instead of J. Geils's "Love-itis"), "Heard It in a Lunch Song" (adapted from the 1977 Marshall Tucker Band hit "Heard It in a Love Song"), and "Lunchtime at the Oasis" (from Maria Muldaur's "Midnight at the Oasis") appeared. When Lushbaugh left, Tony Berardini elevated Tom Couch into the production director's chair and a fresh talent named Eddie Gorodetsky appeared. The new hire was brought into the station and put to work as David Bieber's assistant and a voice talent on

Laquidara's show. "I was working at the *Phoenix*, and across the street was the Rainbow Rib Room," Bieber recalled. "Eddie was a counter guy, a fry cook, and he went to Emerson College. I saw his great wit interacting with the 11:00 p.m. drunks looking for a burger and I said, 'You're going to learn a lot of things in school, but the reality is you're probably gonna learn a lot more if you start getting up at 4:30 in the morning and you give yourself away for free on Charles's show.'"

Not only did Eddie Gorodetsky assume a difficult morning schedule, but he also truly connected with Couch, the two becoming a powerhouse production team. "He was so funny and so quick; what a difference he made," Couch commented.

> The day after Pope John Paul the First died, I got into work and he was going, "Tom, Tom, we gotta do 'Popeless!' We'll take 'Helpless' from Neil Young and cancel out the words. [Sings] There is a town in old Italy, with bishops and cardinals to spare . . . [The chorus was] We are Popeless, Popeless, Popeless, Popeless." It aired only once, then we had to pull it! We got Gilbert and Sullivan's "H.M.S. Pinafore" and put together a two-minute bit, a 'BCN opera, called "The Good Ship 104." Then there was "Wop Music" [created from "Pop Musik" by M]. It ran maybe three times before Senator Lopresti, who represented the North End, called in. He was so upset that we debated him on the "Boston Sunday Review"! Our stance was that it was not racism, we were making fun of people who made fun of Italians.

Couch and Gorodetsky inaugurated a new era of Lunch Songs: the Beach Boys remake of "Catch a Wave" entitled "Eat a Sub (And You're Sitting on Top of the World)," "Another Grilled Cow to Go" (defaced from "Another Brick in the Wall" by Pink Floyd), and a disfigured Go-Go's "Our Lips Are Sealed," renamed "Our Lips Eat Meals."

As 1980 arrived and the staff celebrated nearly a year of relative stability, two station stalwarts left for, of all things, local television. Danny Schechter and Matt Siegel became producer and on-air talent, respectively, of a new overnight show on WCVB Channel 5 called *Five All Night/Live at Night*. As a variety concept that combined political guests and human interest stories with live entertainment, the show became a quick favorite with the vampires of Boston, despite its paltry budget and subsequent bare-bones production technique (often the show was shot with only one camera).

Continued success might have prompted WCVB to open the cash faucet, but all hopes of that occurring were dashed one night during a dramatic live television blunder.

Enlisting the help of Oedipus to gather local musicians and artists, Schechter assembled an exposé on the international and local punk rock phenomenon, an item that always inspired passionate debate. As the edgy show was nearly wrapped without any major slipups, the producer allowed himself a big sigh of relief. Despite a studio filled with black leather, colored hair, safety-pin earrings, and lots of attitude, there had been no serious decency breaches. The closest moment came when Schechter noticed that two of the models in the punk fashion show had merely *painted* their outfits on, but the camera tightened on the pair's faces in the nick of time. Now, with the show ending a few minutes early, Human Sexual Response, the band that had performed earlier, was asked to fill out the rest of the time. The band members responded gleefully with "Butt Fuck," an earthy opus, obviously not cleared for the air. This unvarnished display of punk defiance caught everyone by surprise, with most of the vulgar "piece" broadcast live in all of its unexpurgated glory. Schechter freaked out, praying that the 3:00 a.m. gaffe wouldn't be noticed, but a few complaints managed to filter into the front office at Channel 5 the next day. Very quickly, heads were lopped off and the show cancelled. But Schechter and Siegel managed to trade up: the "News Dissector" transitioned into a distinguished career in network and cable television, while the former WBCN midday jock ended up on WXKS-FM (KISS 108) in Boston, playing that same dance music he had jested about less than two years earlier during his Led Zeppelin disco spoof.

Matt Siegel's departure from WBCN also signaled the beginning of a thirteen-year midday dynasty for his successor. As one of the big guns at WCOZ, Ken Shelton had inflicted major damage on his competition, but with the departure of Clark Smidt, his relationship with Blair Radio's starchy local managers wilted. Shelton had attempted to cross the street from 'COZ to 'BCN with Parenteau two years earlier, but the handshake deal with Klee Dobra fell through at the last minute. Then the jock got a call from Smidt at his new station, WEEI-FM, working a fresh format the program director had pioneered called "soft rock." "The Eagles—without the turkeys," "Joni—without the baloney," "Moody Blues—without the blahs," and "Ronstadt—without wondering what tune just blew by you" were all examples of WEEI's clever ad campaign. Smidt not only offered

Ken Shelton, sharing a laugh with Charles Laquidara, arrives to do middays.
Photo by Roger Gordy.

Shelton a good-paying shift but also made him the station's music director. "I thanked Tommy [Hadges] for everything and gave my two-week notice [at 'COZ]. Meanwhile over at 'BCN, they were ecstatic: Shelton's gone soft!' They wrote me off." But while Tony Berardini could be happy that Shelton wouldn't be directly competing with 'BCN anymore, he failed to anticipate WEEI's rapid rise in the ratings, with not only the expected female audience but also an alarming number of men. By February 1980, the *Boston Globe* reported that in their eighteen- to thirty-four-year-old adult demographic target, WBCN and WCOZ's ratings had "seesawed over the past 15 months, with WEEI-FM now in third place."

Ken Shelton was thrilled to be part of such an unexpected success, but his enthusiasm curdled after a year when it came time to sign a new contract. Despite substantial ratings gains, the salary windfall he expected was not put on the table by the CBS executives who ran WEEI. After negotiations failed to tweak the offer, the DJ opened up to friends and business associates, letting his disappointment become public. This is where fortune once again smiled on Ken Shelton. "I had an angel of good luck hanging over me with all the timing. I got a call from David Bieber, who had been one of my first friends in Boston since 1969. I told him I was down in the dumps. He said, 'Nobody knows this, but Matt Siegel has given notice. I know what you went through with Charlie Kendall and Klee Dobra, but things have changed here.'" Within hours, Shelton sat at a table with Bieber, Tony Berardini, Mike Wiener, and Gerry Carrus, and this time the deal remained solid. "When I came over to 'BCN, I knew a lot of the people there. I was a fan of them, so I stepped right in because I felt I was one of the family," Shelton said. "That's part of the wonderfulness of WBCN: you felt like you were part of the family. And that's how everybody felt who listened to the station—they were all personal friends." Now the workday juggernaut: Charles Laquidara, Ken Shelton, and Mark Parenteau were all firmly in place. Their presence, from six in the morning to six at night, would personify daytime radio in Boston for years to come.

That year, the radio station moved for the third time, out of the Prudential Tower and down Boylston Street to the Fenway area. Slightly more than a right-field foul from the home of the chronically pennant-less Red Sox, the rear of the one-story brick building emptied directly into an outdoor parking lot just across the street from Fenway Park. On game nights the roar of the crowds and their groans of disappointment could be heard easily at 1265 Boylston Street, where curious employees could even sneak a peek over the wall at the stats displayed on the park's jumbo screen. Mark Parenteau remembered the first time he saw the place: "Tony brought me in his Volkswagen bus to look at the property, and as we drove by, I said, 'This is perfect, Tony. We have the baseball stadium over there . . . and a gay bar across the street!'" Kidding aside, Parenteau added, "The move was an empowering thing for the station. It took 'BCN out of that gilded tower in the clouds and put it on the street, in a neighborhood. And it became *the* entertainment neighborhood with the Red Sox right there and the Spit and Metro nightclubs [two blocks] over on Lansdowne Street."

The new building provided plenty more space than the Prudential's submarine-like office space with its one hallway and was renovated in stages. "I was the first voice to broadcast out of the new building because they had the news department move over first," Steve Strick mentioned. "They were still doing carpentry in there. It was comical: saws were going off while I was on the air."

"That entire place was built by the engineering department through the miracle of cocaine," Mark Parenteau laughed. "You'd go over there and their eyes would be open with toothpicks, all wide-eyed and crazy, and they had been there for, like ten days straight." As soon as the air studio had reached a sufficient state of completion, the jocks moved over during a clever live broadcast that originated in the Prudential, and then transferred to Boylston Street via Sam Kopper's Starflect mobile studio, which was driven to 'BCN's new digs as Mark Parenteau handed over his shift to Tracy Roach. Kopper recalled, "We stayed on-air all the way out of the fiftieth floor, down the elevator, out the front of the Pru, and into the bus. We were literally and symbolically coming down from our perch fifty floors above Boston, back to the streets. We fired up the bus and drove slowly up Boylston Street, parade style, blowing the air horn, Bostonians cheering from the sidewalks." Bob Demuth, an engineer who worked with Kopper, maneuvered the big Starfleet rig over to the new location, parking it out in front while scores of cars that had tailed the bus searched for similar spaces.

"Then they came into 1265," Tom Couch remembered, "and we played a tape of my voice with the big MGM fanfare behind it: 'Tracy Roach, this is your new studio!' And she said, 'It's so nice . . . it's so big!'" It *was* big, the largest studio anyone had ever worked in, encompassing enough space to house the control board, record library, a band performance area, and a couch for guests to watch from (and fool around on). The rest of the entire station surrounded this central hub, an axis of amazing activity, and home to WBCN for the next twenty-five years.

One of the most significant moments at WBCN's new Fenway location occurred the same year as the move. Of 8 December 1980, Oedipus recalled, "I was on the air the night John Lennon was shot and I had to read the teletype [bulletin] on the air. It was devastating. Then, I was supposed to go to a commercial. I just couldn't do it, commerce seemed so . . . meaningless. I just started playing music, and friends of mine, musicians, began calling. I told them to come on over, and they started picking records out and playing their favorite Beatles songs." Tony Berardini was watching the Monday night football game between Miami and New England when Howard Cosell made his unforgettable announcement on ABC-TV. "I called Oedipus immediately and agreed with him, 'Do nothing but play Beatles music; take all the commercials off the air.'"

"That's the great thing about radio; it's so immediate," Oedipus added. "It's where people can gravitate and be together, share and hear voices, cry and laugh." Jerry Goodwin followed Oedipus onto the air that terrible night. As a DJ in Detroit in 1964, he had actually emceed for the Beatles at Olympia Stadium and sat backstage with John and Yoko at a benefit for incarcerated activist John Sinclair in 1971. "David Bieber came in, and Bill Kates, my producer, who was a musicologist in his own right, was there. The listeners called in and told me about Lennon's influence [on them], then we hustled to go find out if we had the songs they wanted. It was an amazing four-hour tribute I will never forget."

If rock and roll could be visualized as a burning spirit, then the fire at 1265 Boylston Street was the best place to huddle for warmth and comfort. Listeners wrestled with their emotions live on the radio while 'BCN's employees worked through the same grief. Everyone might not have spent time with John Lennon as had Jerry Goodwin, or interviewed the man as did Mark Parenteau, but all had grown up with him. "I saw the Beatles on Ed Sullivan and I was hooked for life," producer and future music director Marc Miller commented. Miller joined nearly all the members of the WBCN air staff, who hung around the station on 9 December, not knowing really what else to do. Tony Berardini said, "The whole day, we put listeners on the air, played Beatles music, and dropped all of the advertising." Everyone pitched in to construct a broadcast tribute to John Lennon, locating documentary sound clips and recording segments for the special, including a stirring editorial from Danny Schechter, which wasn't completed until seconds before it ran on the air.

"We'd been there all day, and at some point I went to the front of the building and looked out the window," Miller said. "There were all these people out there with candles, all singing."

"There must have been five hundred people, who had been at a candlelight vigil on the Boston Common, who walked over," Tony Berardini remembered. "There were so many standing in front of the station that they blocked Boylston Street. We had been talking about our Lennon retrospective tribute all day; they wanted to hear it, but we had no speakers outside on the building."

"So Tony and I went out there," Miller continued, "and we sort of said [to the crowd], 'We know, we feel it too . . .'"

"The two of us grabbed a couple of boom boxes and stood on the steps of the radio station holding them up in the air so people could hear the special."

WBCN "Soundwaves" poster from 1970 featuring
caricatures of "Mississippi Harold Wilson" (Joe Rogers),
Charles Laquidara, JJ Jackson, Andy Beaubien,
Sam Kopper, and Jim Parry. Poster art by Paul Bernath,
photo of poster by Don Sanford.

WBCN "Stereo Rock" bumper sticker from 1971.
Photo by Dan Beach.

WBCN poster from 1973 designed by the master of macabre illustration Gahan Wilson.
Wilson was hired by the station to do the artwork and traveled from Chicago
to sketch in the studio during a shift by Old Saxophone Joe. By kind permission
of Gahan Wilson. Poster courtesy of the David Bieber archives.

WBCN "Chrome" bumper sticker, early eighties.
From the author's collection.

WBCN "Comic Book" bumper sticker, eighties.
From the author's collection.

The air staff in Fenway Park. Taken from the 1986 WBCN Rock 'n' Roll Calendar.
Photo by Ron Pownall/RockRollPhoto.com.

On the parquet at the old Boston Garden. Taken from the 1988 wbcn Rock 'n' Roll Calendar.

Photo by Ron Pownall/RcckRollPhoto.com.

Wicked Yellow's license plate. Saved by Tank moments before the legendary radio station vehicle plummeted from a crane to her death on the asphalt during WBCN's famous "Drop Era." Courtesy of Tank's personal collection.

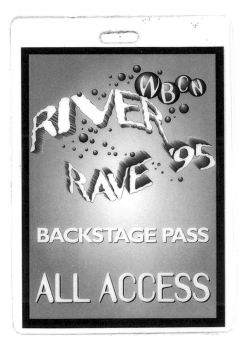

LEFT First WBCN River Rave in 1995, the only one by a river! From the author's collection.
RIGHT 1996 X-Mas Rave All X-Cess Pass. From the author's collection.

The famous eighties period "yellow-winged" and purple WBCN Aerosmith bumper sticker. From the author's collection.

"Rock Revolution" sticker created in 1995 to announce WBCN's new alternative music focus. From the author's collection.

Pearl Jam bumper sticker. From the author's collection.

WBCN/Patriots Super Bowl Champions bumper sticker, 2002.
Courtesy of Tony Berardini/New England Patriots.

Steven Tyler and the author at the Four Seasons Hotel in Boston,
November 2012, talking Aerosmith and WBCN. Photo by Tim Staskiewicz.

Miller shook his head at the memory of the remarkable WBCN moment. "It was one of the most incredible things I've ever seen: all those people just showing up at the station, because they didn't know where else to go."

Later that very same night, Mike Wiener and Gerry Carrus took Berardini out for a previously scheduled business dinner. The pair had already begun working on a plan to leverage WBCN's equity into purchasing more radio properties (they would acquire WXRK-FM and WJIT-AM in New York as well as WYSP-FM in Philadelphia the following year). "They told me, 'We're buying more radio stations; we can't be in Boston all the time, so we want you to be general manager.'" The statement hit the program director like an 18-wheeler. But after the shock passed, Berardini politely declined. "To me, being in the programming end of things was getting away from the music, and I didn't want to get even further away." Wiener and Carrus smiled, finished their meals, paid the check, and bided their time. "They kept asking me, and this went on till the following April. I said, 'I don't know anything about being a general manager. Gerry's comment to me was, 'You didn't know anything about being a program director either.' So, I finally went ahead and said I'd do it."

Berardini immediately chose Oedipus to replace him as program director. His bosses wanted to know why. "I told them, 'Yes, he doesn't know anything about the job, but he's passionate and loves music, and that's what this station is built on."

"Tony took me out to a restaurant to ask me and I totally did not expect it," Oedipus recalled with a smile. "Why me? I was the black sheep!" He pondered through dinner but finally accepted the job and delighted his boss. "Tony said, 'Let's celebrate with some Sambuca!'"

"I ordered us two Sambucas with the beans and all, because I'm Italian, right?" Berardini recalled. "I lit the drinks on fire; we did a little toast and I said, 'Blow out the flame before you drink it.'"

"But when I blew mine out, I blew too hard and it went all over Tony—burning!" Oedipus recounted, laughing. "It was very funny."

"I look up and I saw a wave of fire coming at me! Thank God it went out. I said, 'Great, I offer you a job and you try to immolate me!'"

The new general manager and program director had barely accepted their congratulations before turning to face the surprising reality of WCOZ's second coming. It was entirely unexpected, since WBCN had wrestled its competitor to the ground immediately before and after the strike. In the

WBCN's new location next to Fenway Park at 1265 Boylston Street.
The staff greets Yes guitarist Trevor Rabin. Courtesy of WBCN.

spring of 1980, 'BCN led in listeners with a 6.2 share to 'COZ's 4.1. But by that fall's rating period, which ended in December, the month of John Lennon's murder, a dramatic flip-flop had occurred as 'BCN's share sagged to a 4.2 and WCOZ soared to more than double that with a 9.1. Clearly a great change had occurred at "94 and a half." When Tommy Hadges decided to move on from WCOZ in the summer of 1980, the suits at Blair replaced him with a young programmer out of Phoenix named John Sebastian. During his tenure at KUPD-FM, he had transformed the station from a Top 40 outlet into a full-on rock station. "That was the first station where I used the moniker 'Kick-Ass Rock and Roll,'" Sebastian recalled. The new format specialized in a concentration of rock hits, rotated relentlessly in Top 40 style, without a great deal of sympathy for the eclectic fringes that WBCN so specialized in.

"When 'Kick-Ass' radio signed on at 'COZ, we took a huge hit," Tony Berardini admitted. "It was two hundred rock songs played over and over and over and over again." Sebastian's selections included many mainstream, "meat and potatoes" bands that his heritage competitor largely avoided, like REO Speedwagon, Rush, and Kansas, along with a steady diet of cornerstone rock from Led Zeppelin, the Rolling Stones, The Who, and Jimi Hendrix. He also chose unknown talent if the music sounded right for his format, creating massive hits for newcomers Shooting Star, the Tarney Spencer Band, and Sheriff.

"It pounds home the message 24 hours a day with an endless parade of slogans: 'The Rock 'n' roll Mother.' 'THE Led Zeppelin station in Boston.' 'No Disco.' And the truly hard-sell: 'All rock 'n' roll—no B.S.,'" the *Hartford Courant* reported in 1981. Sebastian ignored criticism and kept up the onslaught. The combination of rock hits with WCOZ's flashy television, billboard, and print ads swept WBCN up in a wave it could not counter. Sebastian offered, "In my opinion, 'BCN believed their press too much, thought they could get away with playing almost anything. They didn't take what I was doing seriously, [so] we caught them unaware."

"We didn't know the significance of 'COZ at that point," Charles Laquidara said. "We just knew that something was wrong." It got worse: in the summer 1981 rating period, "Kick-Ass Rock and Roll" stomped all the way up to an extraordinary 12.6 share, while body slamming 'BCN into the mat with a lowly 4.6.

"It was very clear that if we didn't turn it around, the station would cease to exist," Oedipus stated.

Mike Wiener and Gerry Carrus, who had been involved in plenty of radio battles before this, advised their new general manager and program director to do some music research. Berardini pointed out, "It was the first time that research was ever used at 'BCN. What we got out of it was that listeners thought our playlist was way too broad; we were playing everything, including Rick James and James Brown!" Plus, the station chronically avoided many of the bands that listeners really wanted to hear, simply because the jocks considered them uncool. REO Speedwagon, which had released *Hi-Infidelity*, one of the year's biggest-selling albums, was the prime example. Oedipus said, "We were very, very elitist. In that whole center-stage theory, we weren't even on the stage; we were up in the boxes! We had the talent, definitely more musical knowledge; we were much more creative, but we had to do it in a more focused manner. I had to eliminate some of the musical choices of the DJs because people wanted to hear the hits more often and not hear other songs *too* often. There's a balance there. At the time, I was quoted as saying, 'I'll play the fuckin' Grateful Dead if that's what it takes to play the Clash!' And we did."

In the summer of '81 John Sebastian left WCOZ to start a programming consultancy company that, at its height, advised over two dozen "Kick-Ass Rock and Roll" stations around the country, including WCOZ. Andy Beaubien, the former 'BCN staffer who had crossed over to the competition four years earlier, was promoted to program director under Sebastian's guidance. Beaubien boasted to *Billboard* magazine in the 8 September 1981 issue, "If the Spring book proved anything, it's that WCOZ is here to stay." But WBCN's revamped approach closed the gap; in the Winter '81 survey, 'BCN scored a respectable 5.9, while 'COZ descended from the heights with an assailable 6.7. While much of the reason was WBCN's willingness to adapt its approach to meet the threat head on, the other was that Sebastian rapidly became bored with his achievement, moving on to a new creation he called EOR, Eclectic-Oriented Radio, which would find some success and be a model for the adult alternative radio format. "At the height of [my consulting], I said, 'I've come up with a new idea and if you want to keep doing ["Kick-Ass"], great. It was probably the worst business decision of my life." With the founder abandoning the ship, "there was almost no place to go but down," Beaubien remembered. "When the narrowness of the format began to burn out the music, there was no rescuing it."

"'BCN got a lot better," Sebastian pointed out. "Then they took full advantage of what they already had and what they would never lose: that incredible, legendary image that 'COZ never had. Nothing against Andy, but I think management got too involved and they didn't do the right aggressive moves to stave off 'BCN's rise back."

The tide finally turned in the fall 1982 Arbitron ratings book when WBCN slipped ahead with 5.6 share to 'COZ's 4.9. Not only had the competitor been beaten, but also it would never again rise to challenge 'BCN. As far as me, I was holding down a regular schedule of three weekend shifts when the ratings came out and we discovered that 'BCN had beaten 'COZ. Everyone started yelling and backslapping, hugging, even crying. The tension had been so steady, so complete, for so long that there was this intense rush of jubilation from everyone—jocks to the sales people and even the office staff. The J. Geils Band heard the news and sent over an entire case of Dom Pérignon. Folks were in the hallways swigging the expensive stuff right out of the bottles. This is what victory felt like, and it felt pretty damn good! In 1983, Blair banished the "Kick-Ass Rock and Roll" format, replacing it with an adult contemporary blend of mellower rock sounds. Then, a year after that, even WCOZ's call letters were erased when the station converted to Top 40 WZOU-FM. The deadly "Kick-Ass" Transformer had been toppled and silenced.

I (DON'T) WANT MY MTV

The coolest people in the know, the hippest of the hip, the taste makers, and the trendsetters watched the pitched battle between WBCN and WCOZ unfold with great interest, perhaps even laying the occasional wager on the outcome. But as the boxing match continued through 1981 and into the following year, the attention of these cognoscenti soon dissipated almost entirely. Why bother? To them, the whole radio struggle had suddenly become irrelevant. MTV's debut in New York City on 1 August 1981 represented the opening shot of a brave new world in broadcast media. Even though only a few viewers possessed the cable hardware necessary to receive the twenty-four-hour music-video channel, the prophets anointed the medium as the logical successor to radio, at least the kind of music-oriented radio in which WBCN and WCOZ specialized. In an obvious nod to that premise, MTV debuted with a video clip for the 1979 hit "Video Killed the Radio Star" by the English duo the Buggles. Recalling an earlier era in which the advent of television sidelined radio as the primary medium for mass communication, the song transposed smartly into its new setting. As the cries of "I want my MTV" echoed across more and more subscribers' living rooms, anxieties in the radio business grew. A worry emerged that the whole armature of support for something even as legendary as WBCN would cease to exist, running down like a dying battery, its advertising base vanishing as listeners abandoned the radio dial for their TV remote.

It would be realized in following years that the reports of radio's death were greatly exaggerated; it weathered the introduction of its new competition just fine. In fact, MTV became music radio's great ally because the visual medium more effectively exposed new acts and their sounds to the masses. The songs became three- or four-minute video commercials, introducing not only a tune but also an artist's appearance and manner—the whole package. This made it easier for radio stations to "break" these latest singles on the air and parlay them into hits. As a result, rock music underwent a vast diversification of style as the multihued colors of a new wave scene mingled with the library of tunes from classic bands and artists (at least the ones astute enough to shoot videos). The advent of this sweeping promotional tool helped bolster a sagging U.S. record business, which had hit a recession by the end of the decade. Seasoned warriors like the Rolling Stones and Rod Stewart now advertised their sharp stage moves on the small screen, and a series of creative videos finally pushed the enduring J. Geils Band to number 1 in February 1982 with *Freeze-Frame*. Regional arena-rockers REO Speedwagon, Styx, and Journey all broke nationally with multimillion sellers, while hard rock blasted a path to platinum with AC/DC and Van Halen. Then there were new arrivals like the Go-Go's and their album *Beauty and the Beat*, which was number 1 for six weeks in '82; Joan Jett's chartbusting single "I Love Rock and Roll"; and Rod Stewart soundalike Kim Carnes with her hit "Betty Davis Eyes." Artists of all kinds found their careers jump-started by MTV and then aided and abetted by radio, which steadfastly refused to go away.

A huge part of WBCN's mission, from the first drop of Joe Rogers's needle on "I Feel Free" to 1981's "Turning Japanese" by the Vapors, remained its commitment to the broad palette of rock music. Diversity was a tricky thing to present, but with the turning of the tide against 'COZ, it became evident that Oedipus had gotten a handle on how to do it. He told *Radio and Records* in May '82, "We allowed ourselves to get too far ahead of our audience and we were losing them. We put together a musical structure; I hesitate to call it a format, because the jocks still have freedom of musical choice. But there are boundaries. It's like a painting where the artist works within the boundaries of his canvas. The WBCN canvas covers all the years and types of rock. We play many of the same acts that all AORs [album-oriented rock radio] do; we have to. but there's also other great stuff that should be played that the other AORs won't, everything from reggae to Ray Parker Jr."

Here's an example playlist from a Saturday afternoon shift I did on 24 July 1982 that demonstrates how the presence of some research and formatting had effected the height, breadth, and width of the station's music mix.

Jimi Hendrix, "Stone Free"

The Clash, "Magnificent Seven"

Fischer Z, "So Long"

Survivor, "Eye of the Tiger"

Nazareth, "Love Hurts"

Dave Clark Five, "Because"

Van Halen, "Little Guitars"

Patti Smith, "Because the Night"

Gary U.S. Bonds, "Out of Work"

Steve Miller Band, "Living in the U.S.A."

Ramones, "Do You Remember Rock 'n' Roll Radio?"

Fleetwood Mac, "Hold Me"

Southside Johnny and the Asbury Jukes, "I Don't Want to Go Home"

The Who, "I Can See for Miles"

The Probers (local band from Providence), "Violets Are Blue"

Rolling Stones, "Start Me Up"

Judas Priest, "Electric Eye"

Tom Petty and the Heartbreakers, "I Fought the Law"

By the end of the year, the station would also be featuring the two huge R & B albums of 1982 and '83: Michael Jackson's *Thriller* and *1999* by Prince. We might not have been allowed to play "Flight of the Bumblebee" by

Rimsky-Korsakov, as Charles Laquidara used to, but otherwise, the re-vamping of 'BCN's music policy hadn't become the terrible vise many of us feared. It truly remained "360 degrees of rock and roll."

Since WBCN thrived on new music, the major record labels considered the station a mandatory and essential stop for every band on the road to potential stardom. This early, and often critical, involvement usually endeared 'BCN and its staff to the artists well before they hit it big. For instance, Oedipus's enthusiasm for the Clash placed the band's first two albums high in WBCN's consciousness even though there was little sales evidence in America to warrant the effort. But early airplay of songs like "Julie's in the Drug Squad," "Police and Thieves," and "I Fought the Law" would bear fruit when the band came up with its brilliant double album, *London Calling*, at the end of 1979. Not only did "Lost in the Supermarket," "Clampdown," and the title song become WBCN staples, but also "Train in Vain" went beyond Boston airplay to become a number 23–charting single in the United States with the album a gold seller. After the Clash's finale a few years later, the members continued their friendship with the station: Joe Strummer as a solo artist and Mick Jones as the leader of Big Audio Dynamite. When an unknown Athens, Georgia, group called the B-52's appeared, as colorful as a circus and sporting two female singers each with massive beehive bouffant, 'BCN's airplay of their quirky 1978 independent single, "Rock Lobster," helped the band land a major Warner Brothers contract. As a trio of English bottle blondes, the Police first arrived in Boston in a station wagon, playing the Rat in October of 1978 and releasing a catchy single about a prostitute named "Roxanne." The import 45 garnered an instant reaction at 'BCN, leading to a pair of concert broadcasts from the Paradise Theater in April and the Orpheum Theater on 27 November 1979. The station enjoyed a close association with the Police through the band's demise in 1983, which continued as the members went solo. Over a decade later, the group would chronicle that entire 1979 WBCN Orpheum broadcast when it was released as one half of the *Police Live!* two-CD set.

Then there was U2: perhaps WBCN's most famous band association, along with Led Zeppelin, Aerosmith, and Bruce Springsteen. The legendary story of that group's first appearance at the Paradise Theater on 13 December 1980 has been told and worshipfully retold, each time gaining another layer of mythology, until it hardly seems possible that the tale ever occurred. But, in fact, it was one of those rare, "A Star Is Born" moments, and combined with other exploits in other places, lots of hard work, and a great deal of luck, the members of U2 rode their fantasy all the way to an international superstar reality. The love affair began, like countless others, in the 'BCN music room. U2's Irish import single, "A Day Without Me," had caught my ear in August 1980 at a part-time record store job. I bought the 45, and when I played it for Jimmy Mack, 'BCN's music director, he was totally into it, giving me the go-ahead to feature the song on my weekend overnight shift.

So, I get the credit for the first spins, but things wouldn't have gone much further than that unless two other events combined to create a small "perfect storm." On 20 October, U2 released its first album, *Boy*, in Europe, and one of the few import copies in Boston got into my hands within the week. The album featured U2's brand-new single, "I Will Follow," which seemed even more accessible than the previous one. At the same time, WBCN signed on to host a promotional event for Capitol Records and one of their new acts, a Detroit boogie band named Barooga Bandit, at the Paradise Theater. Capitol would rent the club, and 'BCN would give away discount tickets on the air or sell them on the night of the show. After only a week or so of promotion for the event, U2 suddenly appeared on the bill as warm-up act, an unexpected development given the total dissimilarity of the young Irish band's musical style. Tony Berardini now wanted to bump up 'BCN's exposure of U2 to complement the attention being given Barooga, so I donated my import copy of *Boy* to the station. Within days, the airplay of "I Will Follow," and another cut, "Out of Control," ramped up as all the jocks eagerly featured the songs.

When the evening finally arrived, just before showtime, perhaps 150 people stood about the club as Bono, the Edge, Adam Clayton, and Larry Mullen Jr., all dressed in black, took the stage for the first time in Boston. None of the members were of legal drinking age yet, an innocent bunch with a raw live attack, yet their fledgling set quite impressed the audience, as Edge related in U2's 2006 autobiography: "Our show in Boston was a real surprise for us because we opened for a band called Barooga Bandit in a cramped little club and noticed we were getting a particularly good reaction. We left the stage feeling incredible because the audience was so enthusiastic. Then we went back down to check out Barooga Bandit, only to find that everyone had left. It was then that we realized that they had come to see us."

Sharing the love with U2 from 1980 until the end. The 'BCN staff hangs with the Edge in Foxboro Stadium (1997). Courtesy of WBCN.

The remaining forty people, which included U2 and their manager Paul McGuinness, sympathetic 'BCN staffers, and an embarrassed Capitol Records' representative, stayed to watch Barooga Bandit proceed in an exercise of futility. While that band would soon fade into obscurity, U2's star had just risen above the horizon, leading to many future encounters in Boston, including a pair of headlining sellouts at the Paradise just three months later and the band's first arena show in America at the Worcester Centrum two years after that.

"In the days when we used to come and play the Paradise we made friends with Carter Alan and the folks at WBCN who really supported us. Consequently, our rise in the Boston area was very rapid," Adam Clayton told the *Boston Globe* in 2005.

"WBCN, they were banging U2's music from the very beginning," Larry Mullen Jr. observed in U2's book. "So when we went to Boston, it was a bit like a homecoming. It was a big deal for us. You could say we broke out of Boston."

Bono was more succinct. In a video clip sent to the station for its twenty-fifth birthday celebration, he filmed himself walking along the beach in

Howth, Ireland, smiling broadly, then blurting, "If it wasn't for 'BCN, we'd all be fucked!"

U2 visited the station for the first time in May 1981, for an interview that Ken Shelton found to be one of his most memorable. The members were more than willing to be guest DJs, and the teenage Larry Mullen sounded hilarious when acting the part of a loudmouth American weatherman giving his report. "My favorite interviews were the new, up-and-coming people," Shelton commented, "like U2, who were just so happy to be on the radio." Before their fame had arrived, John Cougar (Mellencamp), R.E.M., Elvis Costello, Nick Lowe, Patti Smith, the Motels, and Talking Heads, among many others, would arrive at the station during this period as young, virtually untested talents. "From Elvis to Elvis," commented David Bieber. "That's what the station was all about."

The experimental attitude was, perhaps, best exemplified by the WBCN Rock 'n' Roll Rumble, an annual event in which the station sought to crown Boston's best up-and-coming band from a field of two dozen contenders. The local competition, spread out over nine nights, was the brainchild of Bieber, who launched a prototype of the contest in 1978 at the Inn-Square Men's Bar in Cambridge. "Eddie Gorodetsky and I came up with the original concept. We didn't want to call it a battle of the bands, and in one of our brainstorming sessions we arrived at the title, which was sort of linked [pun intended] to the Link Wray song [1958's instrumental 'Rumble']." By 1979, the competition had shifted over to the Rat in Kenmore Square for its official first year. "The Boston music community was coming into its own," Bieber continued. "We started out very primitively in 1978, but there was a natural growth and progression of Boston bands that advanced through '79 and '80. We started doing T-shirts, printing supplements [in the *Phoenix*], and putting together consequential prize packages that would be worthwhile to the bands: not just money, but recording time, makeovers, equipment, and advertising space." Various sponsors also climbed on board over the years to offer significant prizes to the winners and also kick in the cash that allowed Bieber much greater clout in placing the station's local advertising for the competition. Ironically, one of those supporting sponsors was the new kid on the block that many people thought would kick WBCN into an early grave: MTV.

The 1979 Rumble awarded the best new talent prize to the popular rock trio the Neighborhoods, while 1980 would crown Pastiche in a battle branded by the local media as a victory for new wave over heavy metal, as the hard-rocking France was defeated. The competition was not without its hiccups: one night a prankster set off a sulfur bomb in the women's bathroom, sending billowing clouds of acrid smoke into the dingy catacombs of the Rat. As bands and patrons alike fled in panic, running madly upstairs into the street, a somewhat-dazed Ken Shelton lingered on, nearly hidden in the smoke at the judges' table, holding his ballot and a couple of remaining free-drink tickets. "Is there a fire?" he asked torpidly, before being shooed out the door. When the Rumble had moved to its new home at Spit and Metro a couple years later, an incident involving a high-powered record label VP threatened to give WBCN a media black eye. The executive, serving as a judge, was caught by a local newspaper reporter out on Lansdowne Street trying his best pickup lines on a pretty lady while the bands toiled inside. WBCN publicly banished the indifferent and intoxicated exec from the competition, and then threw out his scores.

That the writers from such newspapers as the *Boston Globe*, *Herald*, *Phoenix*, *Sweet Potato*, *Boston Rock*, and enduring local rag the *Noise* would devote significant space to covering the Rumble, also serving regularly on the judges panel, ensured tremendous exposure for budding artists who starved for such attention. As the event gathered steam over succeeding years, the music industry sent streams of talent scouts into Boston to size up prospects from each new list of two dozen hopefuls. "For a Boston music fan, it's an opportunity to sample the sounds of the up-and-coming bands," Jim Sullivan reported in the *Boston Globe* in May 1991 (when the Rumble fielded its thirteenth year of competition). "Win or lose, it's a chance to be seen by the record company honchos. That is, after all, the main idea from a band's point of view."

"A lot of our bands went on to get record label deals. 'Til Tuesday was a prime example of that," Mark Parenteau pointed out about 1983's Rumble-winning contestant, which impressed the scouts sent to the competition from Epic Records. Within a year, the quartet, led by singer/songwriter/bassist Aimee Mann, had signed a recording contract and in early 1985 released its debut album, *Voices Carry*, which eventually sold over five hundred thousand copies. After two more 'Til Tuesday albums, Aimee Mann embarked on a solo career that she has sustained ever since. Albert O, who joined the WBCN air staff in July 1982, soon became involved in organizing the annual contest. "Some of the best bands didn't win, ones that all went onto national fame," he observed. "How about the Del Fuegos, Big Dipper, the Lemonheads?" The Del Fuegos, which signed to Slash Records and eventually Warner Brothers, became a much-beloved and influential alternative "roots" band in America for years to come, but in 1983, the group never even won its Rumble semifinal round.

I remember stepping up on stage at Spit at the end of the night to announce the winner. After witnessing four bands, everyone in the audience was well lubed, bellowing out their encouragements and insults, most chanting the name "Del Fuegos!" . . . "Del Fuegos!" . . . Embarrassing emcee situation number 1: the microphone won't work. Well, it didn't, and after a couple of moments of just dying up there, I jumped down and ran through the rowdy crowd into the DJ booth to use that mike. Perched a few steps above the dance floor in a small cupola, the booth might have made it harder for people to see me, but it also provided a greater measure of protection. Good thing, because as soon as I yelled, "The winner is . . . Sex Execs!" the place erupted. Ice cubes were hurled in at the booth, a couple smacking me in the face. That had barely registered before I heard the sound of glass shattering on the wall. The club DJ yelled, "Down! On the floor!" He pulled the door shut and locked it. We huddled in our foxhole and didn't dare poke our heads over the lip until the lights had come on and the bouncers cleared the place out.

By the seventh annual Rumble in 1985, the finals were being held downtown at the Orpheum Theater, illustrating the exploding popularity of the Boston music scene by the middle of the decade. The competition reached a pinnacle of sorts the following year, when the stately old theater was nearly ripped open by the hardcore unit Gang Green. "They were a bunch of punk-surfing, drunken buffoons from Boston," Mark Parenteau laughed, "But, hey, they won!" Front man Chris Doherty totaled a keyboard onstage with a sledgehammer and, for good measure, happily mooned the crowd as he walked off stage, picking up a unanimous triumph from the panel of five judges. David Bieber affirmed the decision: "I think of Gang Green up on that stage, shooting beers," he chuckled, "that band could have been Green Day." The happy and wanton amalgam of misfit toys bashed out their thunderous sonic victory over the more commercial country-rock flavor of Hearts on Fire, whose guitarist, Johnny A, well remembered the night: "It felt almost like Good vs. Evil, because they were such punks and abrasive guys, and we were musicians just trying to write good music. So,

Announcing the 1986 Rumble winners from the Orpheum stage. (From left) Steve Strick, Dan McCloskey, Carter Alan, and Albert O. Photo by Leo Gozbekian.

we lost and, I guess, Evil triumphed!" he laughed. "But, it was a big deal to be playing the Orpheum Theater, and it was a great music scene back then. I miss that." After the finalists had performed, and while the audience waited for the votes to be tallied, the Rumble tradition of presenting a surprise guest artist each year was fulfilled by Peter Wolf and also (the famously nonwinning) Del Fuegos.

As far as sheer attendance numbers went, this would be the high water mark for the Rumble, the finals remaining in the Orpheum for three years before moving back into the nightclub atmosphere of Metro and then the Paradise Theater, the Middle East in Cambridge, and Harper's Ferry in Allston. There would be noteworthy winners who went on to fame: Tribe, the Dresden Dolls, Heretix, and Seka (which changed its name to Strip Mind and featured Godsmack founder Sully Erna), but the central idea always remained the same: focus on the music, no matter what the style and no matter which venue it was presented in. The Rock 'n' Roll Rumble became one of WBCN's most enduring traditions, surviving even beyond the station's FM radio sign-off. In 2011, local music maven and former 'BCN jock Anngelle Wood resurrected the event after a one-year absence, holding the thirty-second contest at TT the Bears in Cambridge under the auspices of classic rock sister station WZLX-FM and the surviving HD-radio and streaming versions of "Free Form 'BCN."

As the decade got into gear and WBCN assumed its new positioning identity as "The Rock of Boston," there were a lot of new faces standing at the bar on those Rumble nights at Spit. Tracy Roach had decided to move on in 1980, and Lisa Karlin, drafted from crosstown 'COZ to do evenings, would remain for two years. The "Duke of Madness" left as well, cut loose by the station after an altercation he had with Charles Laquidara early one morning as the "Big Mattress" crew sleepily arrived to set up the day's show. Neither DJ remembers too much about the argument, but Jerry Goodwin did recall, "There was a wonderful camaraderie amongst the people who worked there, [but] Charles and I were always at each other's throats. I'd rag him all the time about his nonprofessionalism." For years, radio folklore held that the two jocks had a knock-down, drag-out fight in the air studio as the overnight host left and 'BCN's morning maniac took his place. The tale got taller as time went by: one of them choked the other against the studio wall, one flew through the air like that guy in every Western who's tossed through the saloon window in a shower of glass. "That wasn't it at all, man," Goodwin corrected, stating that there were only words between the pair. "It's reasonable to say," he smiled, "that there were professional egos in the way"—that, and perhaps some controlled substances as well.

"It was verbal; no one hit each other," confirmed Bill Kates, Goodwin's producer. "Frankly, it was really a kind of 'tempest in a teacup' as far as I could see."

"Charles said, 'Fuck this!' and walked out one side [of the studio] through one door, and I walked out the other," Goodwin continued. "I lived in Back Bay, so I walked home and went to bed as usual. Then I got that phone call about one o'clock in the afternoon from Oedipus: 'Get to the station right now!' Only then did I realize that Charles had kept walking too and he had not gone on the air. Poor Bill Kates had to take over the show. There were two doors in that studio; we both went out each one, and 'never the twain shall meet.'" Since no licensed operator remained on the station premises, an FCC regulation had been breached, and Goodwin, still signed in on the transmitter log when he abandoned ship, took the fall. Goodwin's departure meant that I finally got a full-time position, as I replaced him in the weeknight 2:00 a.m. slot. Six months later, Oedipus would move me into the 6:00 to 10:00 p.m. shift, which I'd occupy for nearly four years.

A young and boisterous crowd of "weekend warriors," who would take important supporting roles in the subsequent decade, appeared: Dave Wohlman, Carla Raczwyk (known as "Raz on the Radio"), Albert O, Carla Nolin, Carmelita, and Lisa Traxler. New jocks were given their own silver satin WBCN jackets with their name embroidered on the front. "As gaudy as it was," Dave Wohlman remembered, "that was a symbol that you made it; that you were there! What a point of pride: a manifestation that it wasn't a pipe dream, you were actually a part of the staff." The satin would disappear later in the eighties, but not the pride, as Oedipus ordered a new set of custom leather jackets to replace the silvery velvetiness of the seventies model. After four failed audition tapes Brad Huckins finally got his leather prize and was redubbed "Bradley Jay" on the air, a talent destined to be the final DJ on WBCN many years later. One other notable weekend warrior of the time was longtime Boston radio veteran John Garabedian, who joked in 1983 that he was "a semi-legend." But that statement actually rang true: as a star of 1960s AM radio in Worcester at WORC and Boston on WMEX. Garabedian's up-tempo, listener-proactive style would soon endear him to viewers of his own local video channel, V66, and later the hugely successful syndicated Top 40 show *Open House Party*.

Oedipus credited 'BCN's victory over WCOZ to "great announcers, plus great production. We did fun and creative things; production was always an important part of that." From the tone and imagination going into the taping of commercials, radio station IDs, and promos to the "Mighty Lunch Hour" song remakes, WBCN's production department underwent a major generational change as Tom Couch and Eddie Gorodetsky transitioned out after being discovered by Don Novello and relocating to Toronto to write comedy sketches for SCTV. Despite losing this gifted pair, the department didn't skip a beat, owing to the talents that took over. First on the job (some of the time) was Billy West, who would go on to become one of America's great voice talents, following in Mel Blanc's *Looney Tunes* footsteps as Bugs Bunny and Elmer Fudd; originating and starring in *Ren and Stimpy*; acting as a principle voice in *Futurama*; and, heck, he was the red M&M! But Billy West's story was one burdened by incredible personal struggles and years of pain, and to everyone around him at WBCN, each day seemed like it could easily be his last. West, who's been sober for decades, said, "I was just a real bad boy, cross-addicted on cocaine and alcohol . . . high voltage. Once somebody like me started drinking, there was not enough

Billy West before Ren and Stimpy, Bugs Bunny, and the Red M&M! Pictured with (from left) Larry "Chachi" Loprete, Tank, and Peter Wolf. Photo by Mim Michelove.

booze on the street . . . or in the city." But West's hours at WBCN resulted in some of the most memorable comedy bits that the station ever broadcast: "Y-Tel Records presents *The Three Stooges Meet the Beatles!*" and "*Popeye Meets the Beach Boys!*"; "Falk-in' Athol" (the impersonated actor touring the Massachusetts town); "Kangaroo Balls"; "Merry Christmas Boston" (a remake of the Beach Boys holiday classic); "The Rare Elvis Drug Song"; and the fictional, multihour "Fools Parade," allegedly taking place out on Boylston Street during April Fool's Day.

Billy West came to 'BCN through an impromptu, on-the-air routine. "I was living in Weymouth and a friend called me up and said, 'Hey, call 'BCN! They're playing this game, and they want to find somebody who sounds like Mel Blanc.' Charles was up in the Pru, and they were doing "Mishegas," so I called up. Eddie Gorodetsky, who was just a kid then, an intern, was screening the calls. 'Hello, 'BCN. Ya sound like Mel Blanc?' It totally caught me off guard, and I didn't know what to say, so he just hung up on me." West sat there dumbfounded for a second. "Then I said to myself, 'Fuck this guy, I'm gonna call back that radio station.' He goes again, 'Hello, 'BCN, ya sound like Mel Blanc?' And I laid into him with a litany of [Blanc] voices, one after the other, like asteroids hitting him!

"'Wait, hold on.' And he went and got Charles, so I ended up getting on the radio that day." But that was not enough to clinch an on-air berth for West, who was far more into partying than pursuing the opportunity most hopefuls would die for. "I was drunk later that day; I didn't even remember it." As a guitar player, he submerged back into his world of selling instruments and amplifiers in Harvard Square. That might have been the beginning and end of a promising WBCN career, but months or years later (no one quite remembers) a good friend managed to get a cassette of West to Charles Laquidara, who recalled the "Mel Blanc" boy wonder who played "Mishegas" that day. Fate, it seemed, had determined that West would get to the station, no matter what.

Dave Wohlman, who bridged the gap in the production department between Tom Couch's tenure and the arrival of Tom Sandman, mentioned, "I have never worked alongside someone who was so gifted. You had to raise the bar of your own game just to play in Billy's sandbox. If he was working on a character, like Floyd the Barber from *The Andy Griffith Show*, he would talk to you as Floyd the Barber all day! Every day it would change: he'd be Mr. Jinks from the *Pixie and Dixie* cartoons or Larry from the *Three Stooges*." West, as a working musician, could play the guitar and sing in a seemingly endless array of mimicking voices, but his lack of dependability got on Laquidara's nerves. "Billy was just so talented, he was too good," Oedipus mentioned. "Charles would be all upset, and I'd say, 'If he shows up, put him to work, if not, that's just the way it is because we're not firing him! We'll do everything we can do to help him, but we're not getting rid of him.'"

"Oedipus had a very high threshold for my nonsense," West added. "[Once] I didn't come in for two days, and he took me to his office and gave me this stern talking to about being professional. When you're an inebriate, you make whatever excuse [you can]: 'Uh, my alarm didn't go off.' So, at the end of it, I realized he was going to give me a break. He gave me two hundred dollars and said, 'Go get a [new] alarm clock.' I'll never forget it, and we're friends to this day. He knew I was in trouble."

But the problems with Billy West were getting worse. One night he showed up at the station, cornered Lisa Traxler in a production studio, and had to be ejected onto WBCN's front stoop, in the pouring rain. Mark Parenteau, who lived in an apartment across the street and was certainly no stranger to late-night carousing, found West to be a frequent visitor: "Sometimes, he'd get really loud and start knocking pictures off the wall, and eventually even I would have to sleep, so he'd pass out in the lobby of my building. My neighbors would be down there in the morning, stepping around him to get their mail. Billy learned to sleep in the 'Cardboard-only' dumpsters: those were a lot cleaner and didn't have rats." If that sounds like a joke, Parenteau, who started a professional comedy act with West, wasn't laughing. The two began doing a stand-up routine in local clubs, even warming up once for Jay Leno at UNH, and their approach was ad-lib, as Parenteau related: "We were going to prepare lines, but we never got around to it. I was Abbot and he was Costello. I was the straight man who talked to the audience and he'd be . . . Billy!" However, the duo's chemistry failed them at Parenteau's Celebrity Roast at Stitches comedy club in 1984, with West resorting to lowering his pants and making animal noises and grunts onstage. "Everybody was there: Aerosmith, Peter Wolf, Ken Shelton, Lenny Clarke, various writers, Norma Nathan from the *Herald*," Parenteau moaned. "Billy was out of control; it was embarrassing . . . the worst moment." The curmudgeon of chaos had crashed.

"I was empowered by the fact that I couldn't be killed. Every time something came close, I survived it," West remembered. Drug debts loomed, he wasn't paying his rent, and "everybody was chasing me down." West ended up in court frequently, but one visit in his seemingly endless parade of legal tangles proved to be the one that changed everything. "I had to go to court for nonpayment of rent, and there was this crotchety, old Italian judge who looked down his specs at me. He said, 'You never answered a violation from a year ago when you crashed your car on the Mass Pike and left it upside down on the *other side* of the guardrail. You're going to Charles Street [Jail].' Next thing I know I'm in a pissy, smelly, two-hundred-year-old jail, and for the first time in my adult life, I was forced to be sober for seven straight days. Then the radio station was adamant that I go to McLean Hospital, which was rehab. The program was five weeks in-patient." The cold turkey at Charles Street plus West's rehab stuck: "I've been sober ever since; I got it the first time around and never went back. So, if it hadn't been for the station, I'd probably be dead." Upon his return to 'BCN, the dried-out and alert Billy West took all those around him aback: was this really the same person who had clotted up the studios with his dark and drunken presence over the last couple of years? Yes it was, the same man in an astounding transformation, as if a controlling evil spirit had been wrenched from his

Tom Sandman (top left) hosts a WBCN alumni show with (top left to right) Sam Kopper, Jim Parry, (bottom left to right) Joe Rogers, Charles Laquidara, and Al Perry. Yes, Jim is using Al as a saddle. Photo by Dan Beach.

soul and sent, kicking and screaming, back into Hades. Billy West became an unstoppable force in the WBCN production team, now joined by Tom Sandman, and added significantly to an already amazing body of work.

Tom Sandman arrived at WBCN in 1982, after working at the legendary WEBN-FM, in Cincinnati, one of only a handful of radio stations that pre-dated WBCN as a free-form rock outlet. "'BCN was on my radar because there were only four or five stations that I knew had really creative production departments. When I heard those parody songs that Lushbaugh and Couch did, I thought they were the best.'" After the job at WBCN opened, Sandman flew into town, had lunch with Oedipus, hit it off immediately with the program director, and was offered the new gig. "People kept saying to me, 'Wait till you meet Billy!' I had no idea who Billy West was. As it turned out, he was just the greatest partner you could have on a creative level. He was bouncing off the wall with ideas and funny lines. And you know, Oedipus let us do whatever we wanted, he didn't tell us how to do it. He just wanted something new and fresh."

"I hate to use clichés," West said, "but Tom was like Dan Aykroyd and I was like Belushi. Sandman understood music; he could sing and do voices."

"Billy was a terrific vocal arranger; he could hear and mimic harmonies that I couldn't get," Sandman remembered. "For example, the intricate jazz-type harmonies that the Beach Boys would do; he'd get them and then teach them to me." After Sandman worked the music bed out on tape, both he and West would layer on the vocals. "I was the McCartney [voice] and Billy was the Lennon. But Billy had a great David Bowie. I did a pretty good Pete Townshend, [but] neither of us could do Roger Daltrey very well. Billy was a great Mick Jagger; I did an okay Bob Dylan. He did Bruce Springsteen and I did Tom Petty!"

Sandman and West chewed into a new era of lunch songs for Ken Shelton, who gleefully debuted them at noon on the "Mighty Lunch Hour." The lunch song factory turned out spoofs like "Papa Ate a Chicken Bone" (mutated with all its Motown soulfulness from the Temptations' smash "Papa Was a Rolling Stone"); the Rolling Stones' "Mother's Little Helper" became "Hamburger Helper"; and the epic "Flabby Road Medley" was adapted from side 2 of the vinyl *Abbey Road* album from the Beatles. The duo's wit and humor invaded everything the station tasked them to create: from a cheerleading anthem for the Super Bowl–bound Patriots in 1986 entitled "We Love the Pats" to on-air announcements for the Rock 'n' Roll Rumbles or the mountains of commercials assigned by sales—all created in the windowless production studio where Sandman and West barricaded themselves for hours every day. Anyone who visited the station would inevitably find their way to that room to record their station IDs ("Hello, my name is _____, and you're listening to WBCN, Boston").

"Tank would bring in these sports stars, Brooks Robinson and Willie Mays," Sandman remembered reverently. "James Taylor, Steve Miller and Ric Ocasek came in, and David Byrne [of the Talking Heads], who hated to do radio, but he showed up. Jimmy Carter came through, and all these comics, celebrities, and actors: Henny Youngman, Sam Kinnison, and Professor Irwin Corey. Then one day Oedipus walks in with this guy, and he looks vaguely familiar; he's got these kinda' thick glasses and scraggily hair. Oedipus says, 'Tom, I'd like you to meet Andy Warhol.' What the fuck? Andy fuckin' Warhol! How does that happen?" The possibilities in the halls and studios of WBCN seemed endless as the early eighties arrived and advanced. There really was the feeling that anything could, and would, happen. In a magical, multicolored, and electrified atmosphere like that, who really needed their MTV?

NUMBER 1 ROCK 'N' ROLL CONNECTION

Back in 1968, a time that already seemed two thousand light years from home, Ray Riepen's imaginative leap of faith had birthed the successful experiment of underground, free-form radio in Boston. Joe Rogers, Peter Wolf, Al Perry, Sam Kopper, and the other departed soldiers from 'BCN's front line could congratulate themselves for significantly marking a place in radio history; that would have been significant enough. But as the station negotiated tumultuous times, surviving the end of the sixties and the retreat of the counterculture through the seventies, it surged into a new decade in a commanding position, like a marathoner suddenly finding himself at the head of the pack, wondering how he'd ever gained the lead, and now, how he was going to stay there. In August 1985, *Boston* magazine's annual "Best and Worst" poll not only lauded WBCN as the best station in the city but also pointed out that it had been chosen by the Academy of Rock Music as the best in the entire *country*, the decision based on a survey of ten thousand tastemakers in the music business. The article went on to document its success in the wake of WCOZ bowing out of the race: "For the past 3 years 'BCN has dominated the prized Boston radio audience between the ages of 18 and 34, while a clutch of other stations struggle in its wake. Only KISS (WXKS) consistently approaches WBCN's numbers, and it does so with a different, primarily female audience." In January '86, the *Boston Globe* declared, "WBCN remains the station others try to beat," and "the undisputed king of Boston's rock radio." But, although thriving after more than fifteen years of unique evolution through wildly changing times, the station had only just reached the beginning of a whole new role: a prized jewel in a golden corporate crown.

While Tony Berardini and Oedipus focused on running the station and maintaining its fantastic ratings story, Michael Wiener and Gerald Carrus concentrated on building an empire. The pair had begun Infinity Broadcasting in 1972 and then purchased KOME-FM in San Jose and WIVY in Jacksonville before laying out the capital to buy WBCN. Now, they could leverage the tremendous value of their Boston property to finance an expansion of Infinity in earnest. Tony Berardini remembered the moment: "In '81, Mike and Gerry came to me and said, 'We're buying three more stations and we can't run all six [ourselves], so we're hiring a president. You'll like him. He's really good and really smart.' And that's when they hired Mel Karmazin." The reins of power were transferred very quickly to the new chief, and Berardini began to report almost daily to him: "It was [now] Mel's company. I could go talk to Mike and Gerry all I wanted, but at the end of the day, what Mel said, went." A dynamic and intensely energetic thirty-eight-year-old radio sales exec born in Manhattan, Karmazin had worked his way up from the bottom, selling radio ads when he was only seventeen. Soon the rising star was managing radio stations for Metromedia, the broadcasting giant based out of New York City, when he came to the attention of Wiener and Carrus, both mightily impressed with his business acumen and stellar track record. Years later, in 2005, when Karmazin had taken over the reins of Sirius Satellite Radio, Devin Leonard at *Fortune* magazine wrote of the exec's earliest days: "The joke about him was that he was so pushy that advertisers used to buy airtime from Mel just to get him out of their office."

Karmazin became intimately involved in the actions of his small team of Infinity station managers, keeping a tight rein on their bottom line and clearly defining his high expectations. Berardini, at the time only a general manager for a matter of months, vividly remembered the first time Karmazin flew to Boston to meet him. His new boss listened patiently to the endless summary of business data and ratings information presented to him and then issued a simple directive: "Just hit your numbers!" before ending the meeting. Bob Mendelsohn, the general sales manager whom Karmazin appointed to work at WBCN in January 1982, added, "The thing I always remember Mel saying was, 'This is easy; don't make it hard; don't think too much.'" Given absolute authority over Infinity, Karmazin ran the company with a hands-on style, demanding complete honesty, frankness, and performance. "Pain is a great motivator, and one of those phone calls with Mel could be really, really painful if you didn't have the answers," Berardini laughed dryly. "It would be 8:30, 8:15 in the morning, your phone ringing—the inside line, and it was Mel's assistant, Terry, with the four worst words you hated to hear: 'Tony? Terry. Mel's calling.' Your asshole just puckered." Karmazin watched his stations like a hawk, zeroing in on the regular pacing reports to see if each was on target for its projected earnings, but rarely involved himself with the station's programming. "Mel's goals were always based on revenue and profit," Mendelsohn pointed out. "The endgame for him was not about programming; it was about corporate performance." Berardini said, "It got to the point where I knew exactly when the phone was going to ring. Eventually I learned, rather than wait for him to call, which would piss him off, I'd pick up the phone and ask him for help, 'cause if you didn't, he'd assume you had it all together."

"There was lots of pressure on programmers and no slip in the ratings tolerated," Mark Parenteau commented. "Tony and Oedi had to kiss Karmazin's ass. They were under a lot of pressure unbeknownst to the jocks, for a long period of time."

With Karmazin's arrival, Infinity began a rocket ride of acquisition that would astonish the radio industry and Wall Street for two decades and, in the process (for better or worse), pull WBCN out of its singularly unique and isolated space, into a buzzing, interlinked, broadcasting community. "Mike and Gerry tapped into us when they escalated the stakes and started to buy other properties," David Bieber remembered. In November of Karmazin's first year, Infinity bought out three stations from SJR Communications: former disco powerhouse WKTU-FM and WJIT-AM, both in New York City, and WYSP-FM in Philadelphia. The move created instant headlines since the $16 million bill for 'KTU, "marked the highest price tag ever paid for a single radio station at that time," Karmazin told *Billboard*.

"They were acquiring beachfront properties and were willing to spend," Bieber observed. "Then they eclipsed it. Every time they made an acquisition, it exceeded and set a new record [in dollars spent] from the previous time they had made a purchase." A year and a half later, in June 1983, holding to a policy of only buying in America's major radio markets, Infinity absorbed KXYZ-AM in Houston and, by July 1984, had officially acquired the underperforming country-music outlet WJEZ-FM and its AM partner WJJD in Chicago.

Not unlike gunning for Park Place and Boardwalk in a high-stakes game of Monopoly, Karmazin targeted and paid premium prices for A-list radio properties, ones that could be improved (adding little green houses and ultimately a red hotel) and then command the highest returns in airtime sales. Rather than try to breathe life into what was considered a dying dragon, Infinity's programmers "blew up" New York's WKTU's call letters and format in July 1985 and transformed the station into WXRK, known on the air as "92.3 K-Rock." In the same way, market studies indicated that Chicago's WJEZ had hit a dead end playing country music, so the company remade the station into oldies-formatted WJMK. In each case, city by city, as Infinity's empire grew through the eighties, Mel Karmazin's attention focused, not only on acquiring stations, but then also on allowing his team to cultivate them as efficiently as possible. It was a dangerous word: "efficient"; it could change lives overnight, improving or ruining circumstances for hundreds of people who suddenly found themselves invested in, or ex-employees of, an emergent radio kingdom.

Meanwhile, back in the jungle at WBCN, Oedipus was adapting to the arrival of the new Infinity president: "'BCN was responsible for half the cash flow of the company; our success kept Infinity afloat." As long as that success continued, the program director could count on relative noninterference from his corporate mother ship. "He would let us run wild," Oedipus chuckled. "Mel would usually see the logic in it but never tell you it was a good idea. He would just say, 'Okay, I gave you all the rope you wanted; you can hang yourself.'" Michael Wiener told *Boston* magazine in 1985, "They've made WBCN the number-one rated station and the number-one billing station in Boston. Owning 'BCN is like riding a horse: you'd rather have a horse that you have to hold back than an old nag. You can't get anywhere with a dull radio station." So, with nonspecific permission granted, the hairbrained ideas and cockeyed conceptions dreamt up by an inventive staff were coddled and allowed to proceed. No manager would come running, panic stricken, into the studio if Ken Shelton happened to play "Working Class Hero" by John Lennon (with its resplendent and naked f-bombs) or if an occasional "shit" emerged from someone's mouth during an on-air interview. The complete irreverence of a Duane Ingalls Glasscock radio show or any typical Mark Parenteau shift, peppered generously with its crusty humor and blue comments (at least, one of his trademark "Lick me!" exclamations), were left alone and not censored into a sort of radio "white bread" deemed appropriate for mass consumption.

WBCN's front line of talent—Laquidara, Shelton, and Parenteau—occupied the broadcast day, or it might be more appropriate to say that they *inhabited* the day, because that formidable lineup would remain intact and dominate the working hours from six in the morning to six at night, for thirteen years, until Shelton left in 1993. And even after that, the remaining two veterans would stay in their drive-time slots for several more years. "The nucleus of strength at the station was that you had three very distinct and powerful individuals that had significant connection to the community, and ratings that were in the double digits for years," David Bieber commented. "This was kind of unheard of in radio, because in terms of contemporary media people, for the most part, it's all about the bucks. 'BCN could have been seen as a way station on route to Los Angeles or New York. But back then, there was a sense of involvement, a tremendous amount of goodwill between the people, and an ingratiating of talents into this one entity, which was WBCN. And I think at the heart of it, if not the face and the voice, was those three jocks." Certainly, the triumvirate was well compensated for its efforts, and those amounts would only increase over time, but Laquidara, Shelton, and Parenteau clearly *wanted* to stay in their shifts. As the eighties unfolded, the three jocks remained in place for so long that their shift arrangement, on-air personas, and particulars of how each related to the other became familiar and comfortable to even the most casual 'BCN listeners.

Ken Shelton told *The Tab* in 1988 that he considered himself "as the calm between the two knucklehead storms—Charles and Mark." That was true; the less-cluttered, more music-oriented shift became a relatively peaceful interlude amidst the frequent morning and afternoon mayhem. Shelton's image became that of a music sage; the majority of his hours were filled by the sounds of records, not talking. Even so, the midday host recalled that on some occasions the show could get a bit wild: "Toots and the Maytals played in the studio once. He had an entourage of people there just to keep rolling giant spliffs! They were [ripping out] pages of the *Boston Globe* to roll joints and smoke them in the studio. You couldn't see from the studio to the listener line; that's how thick the smoke was!" The interview? "I couldn't understand a word he was saying, and he didn't know what I was saying, but we were all laughing!" Another high point in all those years, according to Shelton, was Bill Murray's June 1981 visit to the "Mighty Lunch Hour," ostensibly to promote his new movie *Stripes* but turning into a full sixty-minute takeover of the airwaves with the comedian reading commercials, answering phone calls, and picking out the tunes. Murray's hilarious finale, a tongue-in-cheek vision of his radio host went like this:

I do I lot of kidding about the way my man Shelton looks, but he's one of the handsomest men in all of the Boston area. I've followed the Red Sox for many years and . . . do you remember when Yastrzemski's kid was a baby? Do you remember what that baby looked like? This is what Ken looks like; he's that beautiful. And he can hit with power, he can throw, he can run, and he can field. Anyone who has money to give him, to promote the American look, the American way of life overseas, should give it to him, and a ticket to anywhere in the world to get him out of this town because he's wasted in this studio, here in this dark room. He just took off his veil and he's beautiful!

A bemused Shelton forced out a "thank you" past all the laughter from his studio guests.

"No, thank *you*!" Murray replied. "How do you keep your skin so fresh?"

Shelton demonstrated his wit and sense of humor often, but his main image was as a music authority. Conversely, Mark Parenteau, also a serious music buff, became much more identified with the growing Boston comedy scene. "Mark was very quick and had an affinity for comedy," Billy West observed. "He didn't want to let that go just because he was playing music [on the air]." It began with the comic slant he gave most of his interviews, when he'd get a laugh from even the most bad-tempered guests, and continued with a short comedy segment he featured every weekday at 5:05 p.m. "Part of the freedom at WBCN, since the beginning, was playing

Mark Parenteau, "The Honorary Dean of Boston Comedy," with Jerry Seinfeld, Adam Sandler, and unidentified Justin Bieber lookalike. Photo by Roger Gordy.

Monty Python, Fireside Theater, Credibility Gap, and all these culturally hippie-dippy comedy bits; they were always part of the library," Parenteau mentioned. "But for my own sense of freedom, or excitement, I latched onto and expanded the comedy format, establishing that, at five after five, we'd play something that was funny. My theory was, people just got out of work, got in their cars, and this was a good way to get their attention. Let's let them know that as they got out of the John Hancock Tower or someplace, and they've been looking at black-and-white figures all day, that the world is still crazy." The daily feature soon gained traction and became a very popular segment, certainly not as famous as "Mishegas," but instantly recognizable with its introduction snagged from a George Carlin album: "Bing! Bong! Five minutes past the big hour of five o'clock!" The DJ frequently shattered his permitted time window for the segment, expanding from five minutes to as much as triple that, especially when comedians dropped in for a personal visit. This led to occasional head-butting sessions with Oedipus, as Parenteau related: "[He] didn't love it because he's a real music guy, but he tolerated it because it got ratings."

Despite the debates over semantics, Oedipus allowed the 5:05 p.m. feature to continue, and the sight of comedians sitting on the front-office couch (laden with its frightening amounts and varieties of DNA), waiting for their cue to perform in the studio, became common. "The Boston comics—Lenny Clarke, Steve Sweeney, Kevin Meaney—a lot of those key guys quickly saw the value of this," Parenteau pointed out. "Suddenly the audience knew them; it was like a band with a hit record." Future stand-up stars like Steven Wright, Dennis Leary, Jay Leno, Jimmy Tingle, Paula Poundstone, Anthony Clarke, and Bobcat Goldthwait all paraded through 'BCN's studio on their way to national fame. The 5:05 p.m. feature became an important catalyst promoting the Boston comedy scene, which now entered a decade of explosive growth, alongside the already-flourishing local music market. "There'd be these little periods where the [music] pickings were slim and comedy was the rock and roll," Billy West mentioned. "It took off! Soon comedy clubs were the place to go." DJ Hazzard, one of the performers who rode the eighties comedy wave, told the *Boston Globe* in 1997, "In the days when things were really good, comedians were like rock 'n' roll stars." Clearly, Mark Parenteau had uncovered a golden nugget, just as Oedipus had exposed the demimonde of a new music scene just a few years earlier. The Boston area's existing comedy clubs began to fill up, and new ones, like Stitches and an improved and expanded Comedy Connection, opened their doors. In 1985, in a direct parallel to the success of WBCN's Rock 'n' Roll Rumble, the station sponsored the first Comedy Riot at Stitches over a five-night span, raising a champion from the two dozen "open mike" hopefuls. Struggling artist Janeane Garofalo, on her way to a long-lived career as a comedian and actress, was one such Comedy Riot winner, as well as Anthony Clark, a young comic going to Emerson College, who would become a star in his own sitcom, *Boston Common*, and other network television projects.

Mark Parenteau, "the honorary dean of Boston comedy," as he was dubbed in WBCN's twentieth-anniversary supplement in the *Boston Phoenix*, developed long-standing friendships with many comics but probably none stronger, or more bizarre, than those with Lenny Clarke and Sam Kinison, both loud and foul-mouthed, type-A performers, who were switched to "On" at birth and never operated below the speed limit. "They could slice you to ribbons," Billy West observed. "Verbally, they could take out your gall bladder and show it to you before you die." Lenny Clarke had run into Kinison in Los Angeles and introduced him to Parenteau, who was immediately taken with the brazen, irreverent style of the former Illinois preacher. "There was no one like him, not only the screaming and all the energy, but, like Lenny Bruce, he said things that nobody said," the DJ remembered. "When there was the world famine going on, and the 'We Are the World' thing, Sam did this routine about Ethiopia: 'Look around you people; what's on the ground? Sand. A hundred years from now it's still gonna be sand. Move to where the fuckin' food is!' It was so politically incorrect, it was hilarious." Kinison found a home on WBCN and visited the station so often that it was hard to remember that he actually came from the West Coast. "Sam told me he loved the fact that 'BCN could and would let him go on and 'say the shit I'm supposed to say!' Every time he came around, as he got bigger and bigger, Sam would do my show and always get mentioned in the *Herald* for some outrageous thing he did or said." The comic became a favorite on MTV for his extreme and very visual persona, also regularly visiting Howard Stern's syndicated show in New York, releasing comedy albums and turning the 1966 Troggs single, "Wild Thing," into a revamped heavy-metal hit.

Smiling, Parenteau remembered,

Lenny and Sam were in Providence one night, and they called up: "We're coming over; you got any blow?" So they showed up at my place at 1:30, maybe 2:00 in the morning. Unbeknownst to me, they had this big Lincoln Town Car and couldn't find anywhere to park, so they decided to pull the car up on the sidewalk, totally in the way of a fire hydrant. Sam had this card that he always traveled with, from when he was a pastor, which you put in the front window by the rearview mirror. Priests use these when making emergency calls, when they're performing the Last Rites; they say "Clergy on Call," and police usually give them the benefit of the doubt. So, they left the Town Car running with the flashers on, the "Clergy on Call" sign in the window, and they came up to "just do a line or two." Five, six hours later, the sun was coming up; we had talked all night, called Robin Williams and woke him up, and it was, like, 8:30 in the morning. Lenny went, "I got to get out of here, and we've got to get that rental car back."

I said, "What rental car?"

"Oh, it's downstairs."

"Where'd you park?"

"Ha-ha-ha, on the sidewalk." So, we went down; the car was actually still there, but it had run out of gas. The blinking lights were fading because the battery was almost dead. So now, Sam was saying, "I've gotta get to New York for a gig!" So, they handed the keys to me, and Lenny said, "Here, you get the car back; I'll pay for everything." So off they went, and here was this big Lincoln over the curb, up on the sidewalk, tank empty, and all my neighbors, uh, looking at me. So I had it towed. That was just Lenny and Sam.

In 1982, the WBCN Comedy Riot was still three years down the road and the Rock 'n' Roll Rumble fielding only its fourth spread of local opponents. With Infinity's presence and Mel Karmazin's tight rein on finances, David Bieber now had to organize himself and his small promotions department somewhat. This he achieved, even though at first glance it seemed impossible for the man, given the perpetual state of chaos and vast accumulation of cultural odds and ends that collected in his office (he still possesses all the pink phone message slips left for him during those years). "One of the great things about 'BCN, especially in that period," he said, "[was] if you could justify something to Mel, Mike, Gerry, or even Tony, ultimately there was a sense of adventure. It wasn't like we were treading in anyone's footsteps; we were doing things that were unique, not only in the market, but elsewhere at comparable stations." Along with the centerpieces Rumble and Riot, WBCN began sponsoring a free lunchtime concert every two weeks with a local or breakout national artist, including the pride of East Boston: the Stompers, who drew a crowd of 1,200 in the middle of a workday. The annual blood drives, on-the-road beach broadcasts with Mark Parenteau, custom station T-shirts available at the "Rock Shop" in local Jordan Marsh locations, a myriad of movie screenings, constant ticket giveaways, and an endless parade of colorful bumper stickers and posters were all part of Bieber's vision. It's useful to note here that what seems conventional for a radio station to do now wasn't necessarily part of the promotional or technical vernacular in 1982. Wiring up Mark Parenteau to do his afternoon show live on the air while he rolled along various MDC beaches in a station vehicle, attracting a cloud of listeners who met him at every stop, was not commonly done in radio. Oedipus or Bieber might have dreamt these ideas up, but the problem-solving arm of engineers (David Stimson and Eddie De la Fuente during this time) made them happen, sometimes only because they brought along extra duct tape or knew the name and inside number of a guy working at the phone company.

Bieber kept his budget down by working with a staff of low-paid (or no-paid) employees and interns. No one was really complaining: they were working for WBCN and loving it. Tank had moved from running the Listener Line to producing "The Big Mattress," doing a morning sports report, and handling errands in the new 'BCN van. "I started driving for the station after Charles's show, delivering stuff, and picking up checks and tapes for the sales department. Then, we realized that the van was this mobile billboard, so at night I'd take it and sit outside at different venues where there were concerts and hand out bumper stickers." Of course, this meant that Tank was now working for WBCN around the clock.

> I remember when they realized I needed help. One night I was outside the BCN-trum (the station's on-air handle for the Worcester Centrum—thought up by DJ Tami Heide), and I'd fallen asleep in the back of the van. Now, I'm notorious: you couldn't wake me up, so the van was there at, like, two in the morning. They had people looking for me at police stations; they thought I'd been arrested. So I woke up, yawning, and it was wicked late. I just drove back [to Boston], thinking nothing of it, and then the police pulled me over on Route 290 to make sure I was okay and that someone hadn't stolen the van!

When Bieber realized that Tank was actually working eighteen to twenty hours *a day*, he found the money to add some help.

One of those new staffers was Larry Loprete (eventually known as "Chachi"), who loved radio from the moment he saw John, Paul, George, and Ringo perform on the Ed Sullivan show: "I knew at a young age that I wanted to work in radio because I sure wasn't talented enough to be a Beatle." Although successful as an employee of Polaroid, Loprete sent a letter to Charles Laquidara expressing a desire to do anything for WBCN. The DJ handed the note over to the Listener Line staff, and soon Loprete got a follow-up call from a young female volunteer.

> She said I could start in a couple of weeks and I'd work the Listener Line after work and on weekends. Then she asked if she could ask me a few questions.
>   "Do you drink alcohol?"
>   "No." To this day I still don't drink, never did.
>   "Do you smoke?"
>   "No, I don't smoke cigarettes."

"Do you smoke pot?"

And I said, "Well, you know . . . I do. I like smoking pot."

"Oh! Can you start tomorrow?"

"Sure!"

"Meet me here at 7:00 p.m. Can you bring some weed?"

Loprete showed up on time for his first shift, shared the wealth, and ended up working straight through the night till seven in the morning. "I was calling all my friends and telling them I was working at WBCN, because this was the station that I listened to and admired; it was so irreverent and the DJs were so compelling. I never stopped listening."

Loprete ended up giving Carla Nolin, one of the "weekend warriors" on the air during that marathon twelve-hour shift, a ride home and soon became her in-studio producer. His first paying job at the station was sweeping the garage floor. "Then Tank hired me to be a van driver. One van, me and Tank. There were no bar or beer nights back then, nothing; we drove the van to concerts." Then in November 1982, there was a serious promotional mix-up at an Aerosmith concert in Worcester, in which twenty-five pairs of tickets were accidently given away *twice* by David Bieber's assistant. Tank and I stood by the van, parked next to the Centrum entrance, conspicuous in our silver satin WBCN jackets. But we would rather have been incognito as several irritated listeners confronted us about the absence of their names at the will-call window. "Hey, Carter Alan . . . hey Tank!" We handed over our own tickets to the first disgruntled winners, but when so many more showed up, we realized this was a major screwup. Thanks only to the station's reputation and close relationships with Aerosmith and the promoter, Don Law, were we able to slip all the ticketless into a sold-out show. It was an embarrassment I never saw repeated at the station. "Twenty-five pairs of tickets that the station didn't have," Loprete reiterated. "I'll never forget it because it gave me my job." The aforementioned assistant was ousted the very next day and Loprete elevated into her position. "Now, I don't smoke much now, but back then, on the day I got hired, I was in the garage in my car with a bag of weed, rolling up a celebratory joint. I saw Tony Berardini, the general manager, coming toward me! I rolled my window down, the pot was on my lap, he saw everything, put his arm in, shook my hand, and said, 'I want to congratulate you on your new job.' Then he walked away. Amazing! It was no big deal back then."

Promotions in motion: Mark Parenteau and Larry "Chachi" Loprete in the "classic" sixties television Batmobile. Photo by Leo Gozbekian.

The "#1 Rock 'n' Roll Connection!" Courtesy of the David Bieber Archives.

For his love of WBCN, Loprete said goodbye to Polaroid, axing a $30,000-a-year job to work as an assistant in a radio station promotions department, a trade he says he never regretted: "Working with David [Bieber] was the greatest experience I ever had. He's a visionary, and I always thought he was the heart, soul, and conscience of the radio station. David did have his idiosyncrasies: he does his best work at three in the morning, so he always stayed up all night and didn't come in till one in the afternoon." That left Loprete the job of managing day-to-day operations of the small department. "He was a real rock to depend on, including his wake-up calls to me every day," Bieber noted about his new assistant. "He was excellent in directing other people as they came along, like Adam Klein and Jason Steinberg. And it was also nice to see his growth and development as a celebrity." Loprete's enthusiasm and wit became obvious to the DJs, who started incorporating him into their shows. "They would need assistance on the air to describe contests and draw winners, so Ken Shelton and Parenteau would call me in to help." He did this so much that he became known as the "Vice President of Prizes," and when newsman Matt Schaffer dubbed him "Chachi" in a quick aside one day, the nickname stuck, and remains to this day. Shelton commented, "Tank and Chachi, at first, were helpers, but they grew into the 'BCN family. The listeners welcomed them: 'Hey! There's Chachi driving the van! Hey Chachi!' Then, a guy who could barely carry on a conversation when we first met him grew into a real talent on the air."

WBCN's on-air slogan became "Your Number 1 Rock and Roll Connection," as the station sought to be top of mind with as many aspects of its listeners' lives as possible. Under David Bieber's direction and Infinity's willingness to sign the checks, the promotions department was given the green light to expand the station's profile and reach. "There had been a point in the beginning of 'BCN where [the station] hadn't been attuned to promotion, marketing, and creative services," Bieber stated. "Over the next ten or fifteen years, there was a logical expansion of those things, because it became important to sales and programming that they all get done." In 1968, any attempt at a promotional tie-in with a client—for instance, allowing Miller Beer to hang its posters at a Boston Tea Party show sponsored by WBCN—would have been viewed as outrageous and improper. But in the eighties, that sort of practice had become an accepted and even necessary way of doing business. Using sponsor money to help generate promotional paraphernalia or foot the bill to produce events became the norm, as evidenced by the flood of WBCN T-shirts and bumper stickers that now accompanied every station benchmark. This promotional relationship was explored and developed early in the decade at the Rumble and then at WBCN movie screenings. "It was unique to the time," Bieber mentioned. "We partnered with the company that had the first-run theaters in town and the suburbs, doing film promotions where Chachi would host [at the theater], welcoming everybody, giving away prizes and creating the 'BCN identity at any screening we did. There was the movie *Caveman* that Ringo was in; we gave away the caveman suit that Ringo wore. The point was to make any film bigger than just a screening." The station's relationship with the local theater chain became so significant that a WBCN introductory film clip, "based on an *ET*–*Star Wars* trailer, ran before every feature in every theater they owned, every day, for years," Bieber added.

WBCN's presence at local concerts proved to be one of the most effective ways to brand the station with the key artists of its core library and new groups on their way to success. Interacting with listeners and the artists themselves at the shows only enhanced WBCN's image as an FM rock station that was on top of what was happening culturally. The most memorable on-air broadcast of this period occurred during David Bowie's visit to Sullivan Stadium in 1983. WBCN had always featured David Bowie's music through every bend in his career, enjoying a strong relationship with the artist and placing the station at the forefront of a long list of those seeking access. Permission was granted for a free-ranging live broadcast from, and during, the event, which would draw over fifty thousand spectators. Along with the broadcast would be an all-encompassing, alphabetical "A to Z" presentation of Bowie's music catalogue on the air. As a resident David Bowie fanatic, Oedipus assigned the lead commentator role to himself, bringing along Bradley Jay, the recently arrived DJ who also knelt in the church of Ziggy Stardust. Practiced and professional are skills one hopes to achieve from years of broadcasting experience, but the excitement of being backstage and close to one of their idols transformed the two jocks into "giddy and happy little girls!" Jay remembered. Albert O, also assigned to the broadcast that day, and working in the audience area and press box, added, "Oedi and Bradley were such rabid fans and they got so excited! That's what 'BCN was all about: we were the fans; we were the spokespeople for all those who were out there listening."

Early in the afternoon, Jay checked in, live on the air, from Sullivan Stadium. He spoke into the microphone in hushed tones, mere yards from David Bowie as the star ate dinner backstage. "Over to my right is a big, long black car, and out of that car came David Bowie and Bianca Jagger. They went over to the buffet line and right now David's about . . ."

"Ten feet, ten feet!" Oedipus piped in excitedly, trying with difficulty to keep his voice down.

"Well, maybe twenty, Oedipus . . ."

"No . . . fifteen!" the program director successfully negotiated.

"Okay. He's eating with Bianca, and we could tell you what he's eating, but that would be kind of tacky . . ."

"Lobster. They're eating lobster!" Oedipus delightfully supplied. Bradley just chuckled before continuing: "Remember Arthur, his bodyguard in *The Man Who Fell to Earth*? Well, he checked our passes; he came over here and it was . . . just amazing . . . and my mind is just fried!"

During the broadcast, throughout the day, the "David Bowie A to Z" continued on WBCN. Jay and Oedipus actually got to talk to Bowie about the alphabetical list when the two were invited into the star's dressing room, a moment that Oedipus hysterically recounted on the air a few moments later. "We were just in David Bowie's dressing room! The rain is pouring and we're in plastic bags," he reported at machine-gun speed.

Bradley gave him the WBCN umbrella so he could go onstage in the rain! Bradley and I . . . sat in the limousine that David Bowie was sitting in, and after we got out of the limousine we went over to David Bowie's dressing room and we talked to David Bowie! We told him we were playing "Round and Round" at that very moment, what you were playing Carter [I was the DJ anchoring things back at the studio], on the air at WBCN, and he was listening and he loved it. We said, "David, are there more than 226 songs that you ever recorded?" And he said, "I don't know." And then Bradley said, "Well, we have 'Liza Jane.'" He said, "Then you have everything I ever recorded!"

With the barest of breaths, Oedipus dove back in at a hundred miles an hour: "It is happening here, the rain is coming down, Bowie's going on in exactly one minute, and it is so intense, everyone's jumping; he can't wait. We are so excited; we shook hands with David Bowie!"

"Twice!" Jay shouted.

"He loves Boston! Back to Carter from the Bowie Connection: WBCN Boston!"

Losing their cool and all vestiges of stodgy professionalism to become an amazed pair of civilians with minds completely blown, both Oedipus and Jay were living the listeners' dream. The moment was infectious . . . and vastly entertaining. But it would become even more so after the concert, when the tag team arrived, spent and happy, back at the WBCN studio to join "Raz on the Radio" as she worked her 10:00 p.m. to 2:00 a.m. shift. Oedipus sidled up to a microphone and began recapping the night: "While Bradley and I were backstage and Albert was up in the booth, Bradley and I snuck into the limousine. Actually, we know Michael the driver, great guy, good friend with David Bieber. We snuck into the limo, and he said, 'David sits on this side and Bianca sits on this side' . . . so, we looked in the ashtrays and what [do] we have here . . ."

"Can you picture this?" an amazed Raz said to her listeners. Here they are with . . ."

". . . a David Bowie cigarette and a Bianca Jagger cigarette," Oedipus finished.

"Why don't you describe the butt itself, Oedipus," Jay chimed in as everyone laughed.

"Bianca's has red lipstick on it; she smokes Carlton. I get that one. And we took two Bowie cigarettes; one goes to Bradley, and he's going to give it to his girlfriend, actually. Bowie only smokes about a quarter of the cigarette; then the rest he stamps out."

"Hold it close to the microphone so we can see it, Oedipus," Raz joked.

"It's a Marlboro, kinda stamped out," he continued. "What we're going to do here, the three of us, Albert, Bradley, and myself, [is] we're going to smoke David Bowie's cigarette. His lips have been on this cigarette."

"Molecules from David Bowie!" Jay emphasized in wonderment.

". . . live on the air!" Oedipus yelled.

Someone produced a match, and the sacred smoking session began. After taking a drag, Bradley Jay observed, "Along with David Bowie's molecules, I get a few of Oedipus's here too!" Out of the litany of voices and laughs, Oedipus asserted, "This is not put on; everyone has to know that we actually sat here and just absorbed David Bowie." Jay loudly inhaled and then blew out smoke next to his microphone, then, passing the smoldering butt over, he said woozily, "Albert, it's your turn; I'm kind of getting dizzy!" The cigarette went around and around in a ritual even David Bowie himself would probably have found hilarious, odd, or maybe a little disturbing if he was listening. As Carla Raczwyk wrapped it up, telling her listeners, "You wouldn't believe it . . . you wouldn't believe it!" and started up another song in the "A to Z," the jabbering excitement of the DJs, having now partaken of some authentic Bowie DNA, could be heard mingling with the opening notes of "Sweet Thing" from *Diamond Dogs*. "It was the greatest show we've ever seen in our entire lives!" Oedipus shouted, before Raz, laughing, shut down the microphones and the music took over.

"The Bowie butt story," Oedipus later recalled, shaking his head; "we were such fans!" The program director revealed that the occasion had an interesting sequel associated with it, involving Bianca Jagger, who returned to Boston, years after the illustrious night of DNA absorption. The celebrity arrived at WBCN for an interview and stepped into Oedipus's office afterward. "I pulled out [her Carlton] cigarette butt and told her the story. So, she signed it . . . and then she left really quickly! 'Okay, that was the program director, right?'"

That was a big part of what WBCN was all about: you were allowed and encouraged to be a fan, champion a band or even go completely bonkers like Oedipus and Bradley Jay. Passion counted . . . and thrived. Nowhere was this more exemplified than in the WBCN "music meeting," where the DJs got to choose the new music that would be played on the air. Jay observed, "The music meetings were part of what made 'BCN what it was. They were both evidence and proof that the jocks really did have some input." The meetings, always held in Oedipus's office, "were not manda-

Bob Kranes, Oedipus, and Tony Berardini at the 1985 Rock and Roll Expo.
Photo by Heidi La Shay.

tory," Jay added. "But, if you were worth your damn salt, you showed up!"
Bob Kranes, who arrived at WBCN in September 1983 from Long Island's
WLIR-FM, to be station music director, told *Radio and Records* magazine,
"All the jocks get copies of records and are invited to the weekly music
meeting, where their vote carries the same weight as Oedipus's, Tony's
and mine. Majority decides whether or not a record goes on the station."
Part-timer Tami Heide remembered, "It could get pretty passionate in
those meetings. I'd stand up or whack a pillow, and someone would yell,
'We *have* to play this!' I remember Oedipus slamming his fist on the desk.
I definitely had opinions and so did everyone else; it was a good group."
Typically, the music director brought in copies of the week's releases that
the record labels were plugging the most, and discussion would ensue over
those tracks. Anyone had the right to comment, reject, support, or rebut
as Oedipus acted as a sort of master of ceremonies. Jay added, "We would
listen to the new songs and comment; of course, we'd try to be funny because we all had giant egos! Everybody could see your personality in there,

so if you were going to tell a joke or something, it was good to make sure it was going to be damn funny," he laughed.

A typical music meeting scenario was as follows: Oedipus sat at his desk, intently considering the piece of paper in front of him with its summary of the songs already on the air and new ones to be considered. A plaster bust of Elvis Presley's head looked over approvingly from a window shelf. Bob Kranes reclined in a chair in front of Oedi's desk, introducing the record company priorities and keeping official score of the meeting's results. The DJs sat around the room, early birds on the soft couch and last minute arrivals on the floor. The tape player or turntable, which was cranked to eleven, warded off unwelcome visitors as song after song got its crucial shot. Based on the evidence of sales locally or in other markets, a buzz on the streets, and the reaction of the group, WBCN would choose its music. "Everybody else [in radio] was really starting to rely on call-out research, but 'BCN always relied on the gut of the guys on the air," Kranes pointed out. "I can't explain it, but we kind of knew what felt right." By the mideighties, this working social gathering, to choose the immediate tactical music approach for the radio station, had become an anomaly in the industry. Very few radio management structures would allow this much rope to hang a station on. Yet, the process worked at WBCN; the jocks invested in the sound and direction of their station like few others. "The jocks were there to personalize it, talk about it, and make it exciting," Kranes added. "The mantra was, you were always afraid *not* to listen to 'BCN because you were afraid you were going to miss something."

So, even though WBCN had come a long way since "The American Revolution" days of It's a Beautiful Day, Lothar and the Hand People, or Zappa's *Chunga's Revenge*, the audience still found the station to be one of the most outrageous, compelling, and addictive entertainment mediums not just in the city but anywhere. WBCN wasn't totally free-form, but with traits like the generous slack allowed each jock on the air, the creativity encouraged at every level in every department, and the presence of open music meetings, it was still pretty free. This could still be said in the early eighties, even as the Infinity Broadcasting logo was added to 'BCN's stationary and Mel Karmazin constantly loomed above, always just one phone call away from changing everything should the station slip or stumble.

CAMELOT REDUX

WBCN's second age of Camelot arrived and set up a decadelong residence at 1265 Boylston Street. In the station's backyard at Fenway Park, Red Sox seasons came and went, including the heartbreaking '86 World Series loss to the Mets (one home run ball shot over the left field wall and dented the station van parked on Lansdowne Street), but 'BCN remained a champion in ratings period after ratings period, year after year. That spectacular success fueled revenue growth that was mostly acceptable to even the voracious Mr. Karmazin, although the pressure on the sales department never relaxed as revenue targets were reached and always increased for the next cycle. The worker bees started referring to that inevitability as the "Mel Tax." "Obviously it wasn't going to Mel directly, but it was an expectation that success was supposed to breed more success," David Bieber explained. "You were expected to expand the boundaries, whether it was promotions, sales, or whatever." While "percentage" and "double-digit gains" were terms that could send Tony Berardini and Bob Mendelsohn diving for the Tums, the sales department nearly always managed to cover Karmazin's spread, so the magical umbrella over programming never folded. To the public, 'BCN remained the coolest thing around, and the DJs were free to frolic in the biggest radio playground in town, perhaps the entire country. To be sure, one of the key attractions of the station was that its listeners, and sometimes even the employees, didn't know what, or who, to expect most of the time.

"Darrell Martinie would come in to record his 'Cosmic Muffin' reports on Tuesday; that was Darrell Day," producer Tom Sandman recalled.

> We found out that Little Richard was coming to the station on a Tuesday [for an interview with Mark Parenteau] and Darrell was so excited: "I gotta meet Richard!" So, he gets there at ten o'clock in the morning, gets all his reports for the week done, and then just sits. He's dressed to the nines and wearing three times as much gold as usual; he can't wait! So, finally at two in the afternoon the studio door opens and Little Richard walks in. Darrell stands right up and says, "Hi, my name is Darrell; they call me the Muffin." And Richard looks at him up and down and says, "Ummmm, hmmmm. I eat muffins for breakfast every day!" Well, they hit it off, and we had about an hour with Little Richard, and he was just the coolest guy in the world. He loved us and we loved him.

Little Richard, the rock 'n' roll star who lit up the charts in the fifties, wouldn't be the kind of personality you'd expect to show up on a rock station in the mideighties, but WBCN was far less a slave to format, and more a creature of lifestyle to its audience. Richard's music had not been featured a whole lot since the days of Peter Wolf and Little Walter, but the perception was that many listeners would still be curious about the effeminate legend-turned-preacher, who used to wear glittering outfits onstage, put red lipstick on in morning, and helped build the musical foundation that WBCN stood on. It was up to Mark Parenteau to make it all work on the air, which, by all reports, he accomplished admirably, resulting in a hilarious and engaging interview. Year later, when Larry Loprete developed a personal relationship with Tony Bennett's family and management, the renowned singer offered to come by the station for an on-air session and "meet 'n' greet" with the staff. What possible common ground did 'BCN have with Tony Bennett? For most, he was a popular singer from a far-distant era with whom they weren't terribly familiar. Was "I Left My Heart in San Francisco" ever on the playlist? But a couple of generations did consider Bennett a major star, and he had made a recent appearance on MTV. Why not acquaint WBCN's sphere of fans with a legend? Nonstop surprise was another reason for BCN's nonstop success.

Since the strike, Tank had developed into a major presence on "The Big Mattress," doing his sports reports and now given freer rein to embellish those short bursts of scores, stats, and schedules with interaction from the athletes themselves, whom he was getting to know personally through his association with the station. He interviewed newly arrived Celtics tender-foot Larry Bird during a live broadcast from the window of the first and, at the time, only Jordans Furniture, on Moody Street in Waltham. "Bird was truly the 'Hick from French Lick,'" he laughed. "Bobby Orr was on the air with us, and I loved local boxing, so we used to have a lot of boxers on: [people like] Johnny Ruiz, who briefly became a world champ, Marvin Hagler, and Micky Ward." Prior to Roger Clemens's rookie season with the Red Sox, "I believe I was the first to interview him in the Boston media, and we became friendly. We had an athlete on once a week during their

The WBCN Sportsrockers with Roger Clemens in Fenway Park. (From left) former Red Sox player and Jimmy Fund representative Mike Andrews, Charles Laquidara, Clemens, Oedipus, Larry "Chachi" Loprete, and Tank. Photo by Mim Michelove.

[particular] season, and these guys all became buddies: Kenny Linsman from the Bruins, Kevin McHale from the Celtics, and Mosi Tatupu from the Patriots. We never paid these guys; they did it because they liked the station and they did it for free."

Very quickly, it would become standard practice for radio and TV outlets to compensate athletes for a season's worth of on-air check-ins. "We felt kind of guilty because Clemens and McHale were doing reports for free even after other stations were paying them," Tank admitted.

> I remember Berardini saying, "Here's a $5,000 check for Kevin McHale, just say 'Thank you, this is for what you've done. Beginning next year, we're going to pay you so much per call' and stuff like that." Tank was giving McHale a ride somewhere and chose that opportunity to hand the check over to the basketball legend. "Well, what am I gonna do with this?" he told me. "I don't want it." So it ended up . . . we happened to drive by the Ronald McDonald House (a home for pediatric cancer patients and their families). [McHale] said, "Go in there; give it to them."

Tank discovered that Roger Clemens had the same attitude: "He didn't want the money either; he said to give it away. I used to take him over to Dana-Farber [Cancer Institute]. He'd go around and sign autographs, and he never wanted any of that known; he did it for himself. He said, 'Don't say a word on the air [about it].'" There was another favorite Clemens memory for Tank:

> When autograph shows began to get popular, Roger didn't want to do them because he hated when the producers of those shows would charge people for his signature. At the time, [athletes] were getting like $200, $300 an hour to do these shows. I said to him, "Well, when those producers come around, just tell them you want $7,500 for three hours! You know, 2,500 bucks an hour should scare them off." So I get a call three weeks later; it's Roger: "You asshole!"
> "What?"
> "Three people said yes [to $7,500], and now I'm committed to doing the shows!"
> "Oops!"

For a time, Judy Carlough, an account executive in the sales department, flexed her knowledge and love of sports by doubling as "Scooter" on the

One of two charity baseball matchups between the excellent Huey Lewis squad and the WBCN Ballbusters. Tank interviews Lewis at Boston University's Nickerson Field. Photo by Mim Michelove.

air, doing the later morning reports for Charles while Tank handled the earlier part of "The Big Mattress." While Carlough's success in sales would prevent her devoting more time to the air, the very fact that she survived and thrived in such a male-dominated media position was way ahead of its time. Like so many "innovations" at WBCN, the presence of sports reporting evolved naturally, even though Tony Berardini and Oedipus never set a specific agenda to tackle it as a programming objective. Tank and Scooter were given great license and encouraged to pursue what started out as their hobby. On 7 July 1986, that hobby became official when Oedipus formally named Tank as the station's first sports director. In addition to his duties on air, the new "boss" was the logical choice to coach an unruly group of rock 'n' roll staffers into the ongoing station softball team, now known as the WBCN Ballbusters. The team played dozens of local businesses and organizations as well as the occasional band, like Huey Lewis and the News, whom the Ballbusters faced in a pair of Jimmy Fund charity games. They split, by the way. Huey took the first game in '84, and then the Ballbusters won the grudge match three years later: 11–10, at Boston University's Nickerson Field in front of five thousand listeners.

Tank's position as sports director put him under the auspices of the WBCN news department, which in the first half of the eighties, was still responsible for an important chunk of each day's programming. Sue Sprecher, Steve Strick, and Lorraine Ballard had all left by 1982, their places filled at various points by Dinah Vaprin (whom Laquidara convinced to return to the station after a four-year absence), Matt Schaffer (the "Culture Vulture," who stepped up from the beloved folk-music WCAS in Cambridge), and Katy Abel (a bright, young idealist who jumped ship to 'BCN after doing news at WBUR-FM). "I had the long shadow of Danny Schechter there," Abel remembered. "He had been such an incredible trailblazer. The purpose of news had been set, and it was more than letting people know that there was a lot of traffic on [Route] 95. It felt like I had this mission: '[How] can I use this time that people, who had gone before me, made such brilliant use of." The purity of that resolve, however, was subject to a steadily changing reality. Programming freedoms at FM radio, in general, had been under attack since the early seventies, in response to pressures from upper management and the marketplace. The rollback of programming liberties in the news department came quickly and manifested themselves blatantly by the early years of the eighties, a decade that, ironically, reenergized the political spirit of the nation. Ronald Reagan's conservatism and foray into Central American affairs, the apartheid flashpoint in South Africa, awareness of political torture in countries around the world, and famine in Ethiopia all reawakened a long-slumbering public concern that had lain largely dormant since the end of America's full-scale involvement in the Vietnam War.

A paradigm shift in how WBCN presented the news was already afoot; a move from what Katy Abel identified as "advocacy journalism"—issue-oriented news reporting to promote a social or political cause—to a more staid and pithy rendition of the day's events. Danny Schechter's original practice, and the news department's process, of reporting an issue then providing great detail to clarify it, gave the listeners a chance to choose their own side, even if WBCN's bias regarding the item was hardly concealed. When the news department had reported on the Vietnam War, it was impossible to listen and not know that the station decried America's escalating police action overseas. When Bill Lichtenstein, with tape deck in hand, raced over to the Common to cover a massive demonstration organized against the U.S. government, it was difficult to miss that WBCN clearly supported the liberal, "people's" position, even as the station explored the story firsthand and in depth. "I wanted to continue that advocacy and creative approach," Abel pointed out, "even though the ground was shifting under my feet and I didn't have quite the same editorial license Schechter had in the seventies." Not only was the structure of news reporting changing, but also the station's liberal leanings were beginning to head butt against the spirit of Reagan America. "The audience was becoming more conservative and the medium more commercialized," Abel summarized. It was not so obvious anymore that 'BCN's audience cared to join in on any leftist rants, so now there was always pressure to slide the station's reporting toward a less controversial and more neutral viewpoint. It's not to say Abel caved in completely under that force, but a position that had been so obvious in earlier years now had to be . . . negotiated.

Dinah Vaprin, known for her resolve and often, in her words, "undiplomatic ways," became the news director in '82 and '83 and handled morning drive news. "On 'The Big Mattress,' we always used to do five- or six-minute [reports], then it got shrunk down to three minutes, then it was two minutes, and then it became one hundred and four seconds. I always went over [the time limit], and this drove Oedipus up the wall. One day he just said, 'That's it. You're out.'"

"I had Tony [Berardini] come in one day," Abel remembered with a grin, "and say, 'Hey Katy, I've got news for you: people are listening to us to hear Aerosmith, not about El Salvador!'"

"Tony and Oedipus would say to me, 'We've been doing these [research] focus groups, and the groups say they listen to 'BCN for the music; they don't care about the news,'" Vaprin added.

"WBCN's diminished news commitment was concurrent with the evolution of the radio station," Oedipus stated matter-of-factly. "Listeners told us, in studies, that they did not listen to WBCN for the news. When they wanted news they went elsewhere. WBCN needed to shed the vestiges of the past, like lengthy newscasts and self-indulgent, free-form music selections that were chasing listeners away. We did not need news reports to keep the spirit of WBCN alive."

When Vaprin exited (again), Abel took the news director reins, faced with the grim prospect of WBCN's shift to a more modernized and efficient attitude toward her department. "It was a transitional era in so many ways," she recalled. "So, while I couldn't do fifteen-minute newscasts anymore, I could still coax Oedipus or Charles or Tony into doing advocacy campaigns." Abel's outrage at the deteriorating situation in South Africa, where the government's system of apartheid remained entrenched with the elimination of civil liberties and opposition leaders jailed, inspired her to propose a day of targeted features to focus attention on the issue. "That was an example of a programming initiative which, in the eighties, was kind of out there; nobody was doing anything like that. It really hearkened to an earlier era in 'BCN's history." The idea captivated Oedipus, who had supported the political stance of many an English punk group, like the Clash or the Buzzcocks, when they performed alongside reggae and soul bands at "Rock Against Racism" concerts. Abel continued, "There were enough people in rock 'n' roll who were saying, 'We hate this shit!' So that was a reason Oedipus could give me the cover the way he did; it was a totally punk thing to do." "Commercial Free for a Free South Africa" was the result: a day devoted to South Africa awareness on 12 November 1985.

Throughout the day from 6:00 a.m. to 6:00 p.m., WBCN featured interviews and news segments from political leaders and also debuted "Sun City," the controversial single put together by Little Steven (Steve Van Zandt) of the E Street Band, aided by 'BCN alumnus, Danny Schechter. The single urged musicians to boycott the popular South African resort town where Western rock bands regularly performed. Steven took time out to be the station's guest, explaining the genesis of "Sun City," which featured a cavalcade of stars including Bruce Springsteen, Lou Reed, Bono, Peter Wolf, Jackson Browne, and many other supporters of the cause. Abel was delighted that the station rolled with her idea and that people embraced it: "I don't believe that someone driving around Hudson in a truck was deeply upset about South Africa as I was, necessarily, but I did get enough feedback from listeners to know that they liked and really respected the fact that we cared so much."

"From a programming standpoint, it stood out from everybody in the market, and it was a good thing to do because, who's for racism?" Tony Berardini pointed out. "My thought was that this was very similar to what we did when John Lennon was assassinated: we went commercial free, and it created a bond between our listeners and the radio station. It seemed like the thing that 'BCN would do."

"To their credit, Tony and Oedi took it to Mel Karmazin, rather than tell me to 'go back to the newsroom, you crazy person!'" Abel laughed. "I don't think that Mel gave a shit about the causes that we were embracing, but he was probably an astute-enough businessman to allow Tony and Oedipus to talk him into cancelling commercials." Well, yes, but not exactly. Berardini lamented,

> I didn't make the connection that there was a slight difference in the amount of money involved in going commercial free in 1980 than there was in 1985. Like, a *lot* of money! The sales department was all up in arms, and my phone rang. It was Mel: "Let me get this straight, you're going commercial free all day?"
>
> "Yeah!"
>
> "How are you going to make up the revenue? If I were the general sales manager, I'd be sitting there saying, 'You screwed me out of my bonus! Now I'm not going to make my numbers because Tony took all the spots off the air!'"

Berardini chuckled at the memory, which certainly at the time wasn't quite so humorous. "One of Mel's phrases was particularly applicable to the situation. He said, 'Tony? If you want me on the plane when it crashes, make sure I'm on it when it takes off!' In other words, 'Give me a heads-up, because I'll help you.' He never told Oedi and me what to do with programming at the station, other than saying, 'Don't lose the license!' So, I told him I'd figure the sales problem out." Commercial rates being set as they were by time slot, the reality was that any spots removed from the daytime hours could not be made up during the overnights; they had to be played in a concurrent position on another day. "So, I went to Oedipus and told him, 'We're doubling the spot load for next week!'" Berardini grinned dryly, "Of course, he was thrilled."

Although wbcn's newscasts were constantly under the knife, the "Boston Sunday Review" (bsr) continued, largely unchanged, for the entire decade. "When I took over the bsr, I recall it being a very free-form show," Matt Schaffer pointed out, "but what I did was set up a structure to it. There were features on the half hour, newscasts at specific times . . . a combination of live [segments] and recorded interviews." Abel often worked in a cohost situation with Schaffer but admitted, "For the most part, it was always his show." When they did collaborate, there was tension: "We were like siblings, fighting a little because we were so different. The dynamic between the two of us was always funny; Matt was all about culture and I was all about politics, and we were constantly butting heads. He'd want to have the *Spenser for Hire* guy on while I wanted to spend an hour talking about trilateralism or something." The pair quickly established themselves as master interviewers, facing an extraordinarily diverse collection of heroes and villains on the show. Schaffer mentioned just some of them: "Robert Ludlum, Sting, Allen Ginsberg cursing on the air, Bette Midler, John Houseman was a fabulous interview, Frank Herbert, Robert B. Parker, Erica Jong, Alex Haley, Clive Barker, Julia Child, and some of the most famous chefs of the day." Schaffer also interviewed Andy Warhol for the BSR: "It was one of the worst interviews of my life!" he moaned. [Warhol] didn't really talk that much. I think that's the kind of person he was; he just took on what people wanted him to be." Sixties icon Timothy Leary also visited Schaffer: "Oh, he was crazy, but smart. He was less a person that dwelt in the past and [more] one who was very much involved in the present. It wasn't like I took him through his life [in the interview]; he came in because he had some sort of consciousness-raising software that he was [introducing]."

"I think we interviewed Ronald Reagan's entire cabinet," Abel estimated. "That's something Matt gets a lot of credit for because he developed relationships with publishers. We got all these big names interviews coming through town: whether it was Alexander Haig, [David] Stockman, or Jimmy Carter. They felt, basically, if you were going to Boston on a book tour, you better do WBCN because they own a certain segment of the audience." Since a lot of these interviews were taped during the week, it was not uncommon for 'BCN staffers to arrive at the station by day to the sight of unfamiliar figures standing about alertly, hidden behind sunglasses, with small coils of wire running out of unobtrusive earpieces and disappearing into conservative dark suits. When one of the two doors into the air studio opened occasionally, perhaps as a Listener Line volunteer walked in or out, a high-decibel shriek of music followed them, buffeting the Secret Service agents as they swept the station for dissidents, bomb-making materials, or perhaps those sensitive files Danny Schechter had "misplaced" back in 1971.

In a significant collaboration between the news and music departments at the station, WBCN signed on to broadcast the daylong "Live Aid" concert from London and Philadelphia in July 1985. Organized by Bob Geldof on the heels of his successful holiday benefit single "Do They Know It's Christmas," the worldwide concert telethon was devoted to famine relief for a starving multitude in Ethiopia. Although the presence and performances from the era's biggest stars (Paul McCartney, Queen, and Elton John) and reunions of past legends (Led Zeppelin, The Who, and Black Sabbath) made the on-air commitment palatable from a programming standpoint, the unvarnished messages of alarm expressed on behalf of the Africans were certainly disturbing. Nevertheless, the day became a momentous and compelling event during which any concerns about ratings points were rendered insignificant as, not only WBCN, but also radio stations around the world, lent their airwaves to help mitigate a human disaster.

I ended up doing commentary from Wembley Stadium for WBCN via telephone, which at the time was not as easy as it sounds. International callers still had to ring up an overseas operator to get through to the States, and since this was way before cell technology had become common, utilizing one of the sparse phone booths in the stadium was necessary. Fortunately, I was able to avoid the long lines by compensating a janitor (a trusting bloke) to use the telephone in his office. Thanks to U2, I had a fantastic seat in the press section, about a third of the way back from the stage. Just before the concert began with a performance from Status Quo, the entire Wembley audience suddenly wheeled in my direction to look at . . . me? "Hey everyone, wave to the visiting Yank!" I basked in the fantasy for two seconds and then realized that Prince Charles and Lady Diana had arrived and taken their seats about forty feet away. Of course, every time I tell the story, they get a couple feet closer.

The effort to decry racism, begun with "Commercial Free for a Free South Africa" in 1985, became the last, great advocacy campaign from the news department at WBCN. It continued into 1988, when the station made national headlines with "Shellshock," a monthslong protest over Shell Oil's continuing practice of doing business with South Africa. Charles Laquidara asked listeners to cut up their Shell Oil credit cards and send them in to him as a protest against apartheid. "Let's get the Shell out of South Africa!" became the rallying cry. For nearly half a year the crusade continued, and even though Shell did not stop doing business with the apartheid government, "Shellshock" achieved a great deal by adding numbers to the voices of protest. However, even with this encouragement, the process of paring down the percentage of news on WBCN continued. "This transition was occurring as part of a much bigger transition in the industry," Abel observed without bitterness. "I don't think you can point fingers at individuals at the station who were somehow unable to hold back the tide." There were, however, a couple of big hurrahs left for the news department, as Sherman Whitman, hired five years earlier as a reporter for "Big Mattress," took over many of Abel's duties when she exited in early 1989. It would be Whitman's pleasure to see the dramatic completion of the work Abel had fostered but also his sadness to witness the end of WBCN's legacy as a game changer in the area of local news reporting.

WBCN traveled through the eighties with a couple of powerful positioning slogans: "The Rock of Boston" and "Your Number 1 Rock and Roll Connection," the latter dispensed in dozens of variations such as "Your Number 1 Concert Connection," "Your Aerosmith Connection," "Your Springsteen Connection," and so forth. Ironically, it was Clark Smidt, WBCN's former adversary at WCOZ, who coined the phrase "Your Number 1 Rock and Roll Connection." "I became a broadcast advisor/consultant at the time, and Tony hired me for 1,500 bucks to work on a slogan. That's what I narrowed it down to and they loved it." The image of a Fender Stratocaster headstock and part of its severed neck, sometimes with strings dangling, became the iconic logo of the station throughout the eighties. Plastered liberally on billboards, on bumper stickers, and in magazines, the use of the guitar indisputably identified WBCN as Boston's home for rock and roll. Since the image also showed that an instrument had been shattered, or was in the process of being destroyed, it implied that the station lived its life boldly, like a guitar poised in midflight over Pete Townshend's head rather than one hanging safe and sound on the wall of some music shop. With that in mind, many of the ideas that emerged from the WBCN promotions department around this time involved attitude, altitude, gravity, and motion.

David Bieber and assistant Larry "Chachi" Loprete built their department into a formidable instrument through a thick Rolodex of contacts, an informal "think tank" atmosphere and the spirit of adventure. "We had weekly promotions meetings, and we did go for outrageous things," Bieber recalled. The idea emerged from one of those brainstorming sessions to hold a chancy, but certainly ear-catching, event in conjunction with Elektra Records to promote the newest release from the Cars. "We put Mark Parenteau in a car, suspended by a crane, over the Charles River. He did a live broadcast from up there and [then] we gave away the car. We did not clear it with OSHA or anyone else; we just rolled the dice."

"The wireless mike really gave us the freedom to try stuff like that,"

Parenteau recalled, with a laugh. "The car sat on a platform of wood with cables attached to the four corners; then the crane hoisted us up. I was sure that if the wind picked up enough to cock me at a forty-five-degree angle, the car would just roll off the platform like a bunch of tacos off a plate!" Another hairbrained idea led to Parenteau broadcasting live while riding the legendary wooden rollercoaster at Paragon Park on Nantasket Beach (before it closed down in 1984). "They duct-taped the wireless mike to my arm so I could hold onto the coaster car while it was flying down the track. I just remember getting to the top of the big drop-off, and when we dropped, I began yelling, what I think is, the longest expletive ever broadcast on the radio anywhere: 'FFFFFUUUUUUUUUUU . . .'" Parenteau chuckled, then added, "But, you know, that might have actually been legal, because I never completed the *c* or the *k*; it was just one long "FFFFFUUUUUUUUUUU . . .' while the car was coming down that hill!"

Although 'BCN's afternoon DJ had not been forced to escape a car that was rapidly sinking into the (still highly polluted) Charles River, nor been tossed hundreds of feet from the "Giant Coaster" on the South Shore, the promotions department did begin dropping other objects intentionally. Larry Loprete said, "This was like the early days of David Letterman's show, when he would drop stuff off his roof."

"We would get a tie-in with a client to do a live broadcast," Mark Parenteau explained, "and people would come out of the woodwork to see a five-hundred pumpkin slam into the ground!"

"At Ernie Boch's on Route 1, a helicopter flew in, picked up a giant pumpkin, and dropped it out of a net," Loprete laughed. "Try to do that today with the FAA! Then, the Norwood Chief of Police called us saying that we'd never be allowed to drop anything in the city again because all of Route 1 had been totally backed up. Ernie Boch, obviously, loved it!" The pumpkin drops typified WBCN's penchant for the extreme, and in an overinflated, multicolored "I Want My MTV" era with big, poofy hairstyles, zillion-zippered parachute pants, pompous shoulder pads, *Flashdance* leg warmers, and heavy-metal makeup, "outrageous" scored points. But was it that much different from packing a lunch to go see wing-walking barnstormers at the local airport back in the twenties? People have always loved spectacle, and this was about as big and unusual (and stupid) as it got. Hundreds, even thousands, of WBCN listeners and curious onlookers would be swept up in the delirious moment, excitedly watching an approaching helicopter

Dropping the Great Pumpkin! Parenteau and Chachi ponder the impending pulpy massacre. Photo by Leo Gozbekian.

or a fully extended crane lift its doomed cargo to the heavens. There was the obligatory on-air buildup . . . wait for it, wait for it . . . the release . . . the booming "Thud" and "Splat!" . . . then the perimeter would collapse as dozens of people rushed toward the pulpy mass, reaching in and tugging out armfuls of the gooey innards. As Mark Parenteau ad-libbed the play-by-play for each vegetable assassination, listeners at home or work couldn't possibly step away from the radio. Why? Just because it was a 'BCN happening, and who did that sort of thing?

The idea of gravity as a promotional tool expanded. "It became the 'Drop Era' of 'BCN," Mark Parenteau chuckled.

"Oedipus said, 'Let's drop a piano,'" Loprete recalled. "I don't know why we did it; maybe it had something to do with Elton John. We went through several locations and it turned into this weekly thing: 'Where are we going to drop the piano?' because each city [we approached] turned us down. The newspapers kept covering it, though, and it became the great piano fiasco. Finally, we dropped it at the Charlie Horse in West Bridgewater."

"I can still remember those final piano chords in my head," Parenteau chuckled. "It was great, better than the end of the Beatles 'A Day in the Life.'"

After warming up on titanic pumpkins and the notorious baby grand, the idea to drop "Wicked Yella," the WBCN van, was a natural. The vehicle had accumulated well over a hundred thousand miles on its odometer by the end of the decade and was considered a rattling deathtrap after years of faithful service. Tank, for one, was not sorry to see the gaudy, lemon-toned van, with its emblazoned station logo and image of the Sunbeam Girl decorating the side, slated for a final plummet to glory: "That thing used to break down all the time! We used to have to get jumps from the WAAF van in Worcester, and how embarrassing was that?"

"It was a big Chevy Econoline," Parenteau remembered. "We took the gas tank out of it and hoisted it up by crane to over three stories high in Norwood. It smashed into the ground with this huge crash and made great radio and [also] great visuals for the newspapers." To complete the demolition, a three-thousand-pound concrete block was lifted high into the air and dropped squarely onto the van's roof, before hundreds of spectators rushed in to claim souvenirs.

"They were just peeling parts off," Larry Loprete recalled, shaking his head. "I saw one guy walking away with the driver's side door!"

"The van drop was one of my fond memories," Tami Heide reminisced. "But I remember getting a hard time from some listeners because it was a Chevy, and then we got a Toyota as a replacement. 'U.S.A.! U.S.A.! What are you guys doing?'"

Perhaps the most famous drop cooked up by the 'BCN promotions department was a pair of ingenious re-creations of the 1978 classic Thanksgiving episode of "WKRP in Cincinnati." In that renowned television sitcom, staffers at the hapless radio station set up what they thought was a tremendous promotion to drop live holiday turkeys from a helicopter to hungry Ohioans on the streets below. Unfortunately, no one realized, or they forgot, that turkeys possessed only limited flying capabilities, and the poor birds plummeted, one by one, to their deaths. WKRP newsman Les Nessman (actor Richard Sanders) described the action on the air and his cries of excitement quickly turned to ones of alarm, and then panic: "Oh, the humanity!" mixed with his very visual description of turkeys hitting the pavement "like sacks of wet cement."

"So we hired Les Nessman," Loprete recalled. "We made it sound like we were dropping turkeys, but they were really just hundreds of paper 'turkeys' that my interns and I stapled gift certificates to." The helicopter hovered loudly over thousands of listeners at North Shore Mall, while Sanders, as Les Nessman, began his on-air report and 1,500 paper turkeys were shoved out of the chopper. The resourceful promotion made holiday headlines all around Boston and prompted David Bieber to schedule a repeat of the event in 1990 at the Channel nightclub in Boston. Once again, Sanders reprised his WKRP role while the small paper packages fluttered slowly down to the parking lot below, although this time, many of the "turkeys" were caught up in the breeze and landed in the oily waters next to the club. Loprete remembered, "I couldn't believe it, but the Animal Rescue League called to complain that we were dropping live turkeys. Amazing!"

Meanwhile, back at 1265 Boylston Street, the hijinks continued. When Hurricane Gloria roared up the Eastern Seaboard in September 1985, Mark Parenteau could only think about how to use the storm to his advantage. Since the eye of the storm would pass directly over Boston during his shift, he proposed a live broadcast using the wireless microphone out in front of the station. While the winds howled, tossing newspapers about and rolling wine bottles loudly down the street, Parenteau and a small retinue of DJs and station staffers interviewed a cop in his police van, a brave cabbie

"Oh, the humanity!" Les Nessman and Tank look to see if turkeys really do fly. Photo by Roger Gordy.

looking for desperate riders, and the crew at the convenience store across the street (determining which items had been panic bought by residents of the Fenway). At several points, a sudden and bashing eighty-miles-per-hour wind gust would pitch the whole group off its feet. That's when everyone grabbed onto the weighty Tank, who was, fortunately, a part of the crew. Parenteau might have been the host, "but I was the anchor!" Tank laughed. Not to be outdone, Charles Laquidara, whose colorful life could have gone on privately, instead allowed it to become daily fuel for many of the best bits on "The Big Mattress." The show received no small amount of mileage from the DJ's regular battles with technology. Larry Bruce, who served as the station's custodian and an engineer during the eighties, commented, "Charles embraced technology like no one else I ever knew. He had a full-blown sound system integrated into his house, a backup generator, geothermal wells, the first cell phones, whatever; Charles had it. He gave a shout out on the air to someone one day and ended up with a BMW with a nitrous kit in it. They told him, 'It's not like the old days; you don't pump the gas pedal to start it. Every time you put your foot to the floor, you'll trip a microswitch that gives the car a burst of nitrous.' So, after his show, he went down to start the car, pumped it twice and *boom!* He sent the intake through the hood!" Bruce howled at the memory. "He thought somebody put a bomb in his car!"

David Bieber and his department rolled out the two-day Rock 'n' Roll Expos, held at the Bayside Expo Center for three consecutive years beginning in 1984. "The expos were meant to be a showcase for all things 'BCN, including advertisers in widely divergent booths, sponsors, events, and activities," mentioned Bieber. "The magnet was 'BCN and the music." The DJs sat at long tables for autograph and photo sessions with fans; groups including Cheap Trick, Meatloaf, Joe Perry, the Alarm, and Greg Kihn performed, and even marathon cutting sessions by local hairdresser Jan Bell went on for hours. Twenty-five thousand crowded into the building the first year, thirty thousand the next. "They had to shut down the exit from the expressway because there [were] so many people trying to get to the Expo Center," Loprete recalled.

"That was the point, I think, where 'BCN significantly transitioned from being this music and pop-culture station to [being] lifestyle oriented," Bieber pointed out. "It was reflective of the audience. People were getting married, having children, buying houses, changing their direction. Not to say they weren't going out to concerts and clubs, but it was one of those periods of . . . acquisition, where people were also interested in what Jordans Furniture had to show at the Expo." As such, the scope of station events began to be widened to include all aspects of the listeners' lives: the annual blood drive with the Red Cross; the WBCN 10K Road Race; fireworks over Boston Harbor with an accompanying soundtrack broadcast live over the air; "Row Row to Revere," a canoe voyage across Boston Harbor from Nahant to raise money to fight spina bifida; the WBCN ski team; business card drawings for the professional community; and food drives for the homeless.

Bieber found that just about any crackpot scheme his "think tank" came up with was rubber-stamped by the music business: "In the eighties, the record companies were just awash with cash, especially when CDs came along. If you could think of it, you could secure it from the labels, like 'Have Lunch with The Who in London.' The effort was made to not repeat ourselves and try to give people something they could not buy, an experience they could brag about." Whether it was escorting listeners to the LA Forum to see Genesis or living out a Geffen Records–inspired fantasy to meet the Black Crowes backstage in Denver and then ski all day at Keystone Resort, the wallets were open. "WBCN and Atco Records sent me to Australia with [contest] winners to see this band Goanna," Albert O recalled. "We went all the way to Sydney and stayed there for eight days; it was truly amazing."

Epic Records brought Stevie Ray Vaughan and Double Trouble to Boston in February 1987 and handed over all 1,250 tickets for the Austin rocker's show at the Metro to WBCN. Since the guitarist had played the much-larger Orpheum Theater the previous year, this was a special, more intimate visit, and all the tickets were given away on the air. Vaughan and his band also dined with the 'BCN jocks at Davio's Restaurant on Newbury Street, where owner Steve DiFillippo had fostered a reputation for fine Italian food and impeccable service. Ken Shelton sat at the table with Vaughan that night: "He had his manager call over to Davio's [and ask], 'I wonder if your chef could whip up some rabbit?'" There was no doubt that the first-rate eating establishment possessed a splendid and well-varied menu, but like most New England restaurants, hare wasn't on it. To Shelton's amazement, though, the guitarist's crew had prepared for such an eventuality: "Vaughan's roadie had a case of fully skinned, fully cleaned rabbits with him! And the restaurant didn't even blink; they prepared it! Now that's a true Italian chef, not just your usual spaghetti and lasagna guy."

Stevie Ray and Double Trouble digesting with Oedipus and
Carter Alan at Davio's. WBCN displays major mojo by giving away
all the tickets for the band's 1987 show at the Metro.
Photo by Ron Pownall/RockRollPhoto.com.

The air staff up close and personal with Aerosmith at the free WBCN show in the
Worcester Centrum, 1986. Happy eighteenth birthday to the Rock of Boston!
Photo by Ron Pownall/RockRollPhoto.com.

Perhaps the foremost benchmark of this, WBCN's second age of Camelot, occurred in 1986 when the station celebrated its eighteenth birthday by giving away all thirteen thousand tickets to a free concert with Aerosmith at the Worcester Centrum. The band had gone through some hard times by the end of the seventies as the members split into factions, largely as a result of prolonged (and well-documented) drug abuse, but had reassembled with a new label, a fresh resolve, and an album entitled *Done with Mirrors*. Now on the comeback trail, Aerosmith had never lost its longtime connection to WBCN, which could be traced all the way back to Maxanne Sartori's enthusiastic support; plus, Joe Perry and the band's manager Tim Collins shared a strong personal friendship with Mark Parenteau. "We always had the first link to Aerosmith," the DJ mentioned. "I would always get the spontaneous interviews with Steven, my name was always in the credits, and 'BCN would get first knowledge of any concerts they were going to do. I know a couple of other stations wanted to be a part of all that, but for sure, we were the favorite." When the idea materialized for the band to perform a special show to commemorate 'BCN's long history and legendary status, the group eagerly agreed.

In the official WBCN press release, Joe Perry stated, "The free WBCN show is another gift to the people from Aerosmith, the people's band." Steven Tyler added, "The buzz on this concert is like lending the family parrot to the town crier. The excitement we feel about this show is phenomenal. You can't buy it or bottle it. It's a pleasure to be making history with WBCN." The concert itself was a dizzying peak of excitement for all the 'BCN jocks who assembled onstage before Aerosmith "let their music do the talking." Once they had received the keys to the city of Worcester in a small official ceremony, each member of the lineup was introduced to the audience and recognized individually by Charles Laquidara; then, a memorable photo shoot caught them all in front of the cheering crowd. Later, during Aerosmith's set, the band invited everyone back onstage for a rousing version of "Happy Birthday," while candles spelling out W-B-C-N burned brightly on a gigantic cake. It was a moving moment for everyone gathered on that stage, and even the veterans—Laquidara, Shelton, Parenteau, and Berardini—knew that this moment was as good as it could ever get.

FROM BOYLSTON STREET TO WALL STREET

By the mideighties, the Fenway Park–facing, rear garage-door entrance to WBCN had seen thousands of arrivals and departures, carefully timed to avoid the unsportsmanlike panic of Red Sox Nation traffic. Access from the dank garage and up the dangerously pitched stairs to the offices and studios above took one past the graffiti wall, where employees and stars were encouraged to spray paint their names and messages to posterity. "That graffiti said a lot about what 'BCN was," weekender Lisa Traxler mentioned. "Everybody added to that." The freedom of expression, whether directed toward a small quadrant of cement-block wall with aerosol paint can in hand or aimed heavenward through a microphone and turntable (or the newfangled compact disc player that had just shown up), were one in the same. "I liked Oedipus's analogy," Tami Heide remembered: "All the jocks at 'BCN have the same brushes, the same paint, and they're given the same canvas. But they all have their own brushstrokes; it's just how they put it together." Protected in their studio, the DJs labored over their portraits, even as forces arrived that threatened to limit whatever freedom had been carefully earned during two decades of survival. Challenges loomed from inside the rapidly expanding Infinity empire, as well as outside from a cloud of seasoned competitors and new players in the market.

Infinity Broadcasting had binged on acquiring stations, leaving the company cash poor by 1986. Subsequently, Michael Wiener, Gerald Carrus, and Mel Karmazin took the company public, offering up its common stock to shareholders and raising millions in capital. They used the money primarily to reduce Infinity's long-term debt, which had been accrued when borrowing to buy stations, and to provide funds for accumulating additional properties. Within a year of the public offering, Karmazin went on to purchase six more stations: KROQ-FM in Los Angeles, WJFK-FM in Washington, D.C., and two AM/FM combos in Tampa and Dallas. Even with the sale of Infinity's Jacksonville property and a pair of San Diego stations in December 1986, it was a dizzying expansion. By the time of the company's annual report to its shareholders in spring '87, the number crunchers could claim that "Infinity is the nation's largest, publicly traded, *radio-only* owner and operator of stations." The report further revealed that Infinity had taken in "$1.4 billion in radio advertising revenues (approximately 20% of the total $7 billion spent for radio advertising in the United States)." Although just one part of a massive and highly valuable corporate entity now stretched coast to coast, WBCN was certainly not a diminished commodity; in June 1988, the *Boston Phoenix* disclosed, "The station that Infinity picked up for $3.5 million in '79 is valued today at an estimated $75 million to $80 million."

"I really didn't understand the finance part of it and what it meant to be [a publicly traded company]," Tony Berardini recalled. "What it meant on a day-to-day basis at WBCN was that there was a lot more pressure for us to make our numbers; things became more intense. Being a public company that now reported its earnings quarterly, Mel had to stand up in front of the stockholders and answer questions, and God help you if Mel had to defend your sorry ass."

Bob Mendelsohn remembered, "There were these quick outbursts [from Mel], not necessarily from anger: 'Here's what I need and I'm not getting it! How are we going to keep our stock price up? How are going to keep to our acquisition plan without generating revenue?' It was always a highwire act."

Berardini added, "We were in a meeting one time with a general sales manager who wasn't making his budget, and he was saying, 'Well the ratings are this and the business cycle is bad . . .' Mel says, 'Hey, listen! Don't sell spots, then. Go stand on the front steps with some pencils. Sell the pencils and book that as revenue. I don't care how you make it!'"

"Mel Karmazin was never part of the 'BCN culture, never part of the counterculture; he was a money man," Katy Abel commented. "I mean, nobody said Mel was a teddy bear, but he was a genius at making money, and if that's what mattered to you, then he was your guy."

"Once they gave Mel Karmazin the job of making the most money in the shortest amount of time with the fewest amount of people, the bottom line was the bottom line," Charles Laquidara summarized.

The business world learned to respect Karmazin and his upstart Infinity Broadcasting Corporation, and even though the company still carried an enormous amount of debt with its recent station acquisitions, the amount of revenue its properties generated grew. Tony Berardini revealed that WBCN, the station whose equity had initially financed the growth of the empire, posted 14 percent of Boston's entire radio market revenues in 1986. It was a phenomenal figure, considering that more than two dozen radio signals crowded the city's airwaves. With numbers like these, the future of Infinity seemed only positive at the New York Stock Exchange where its shares were traded daily. That rosy outlook held firm until "Black Monday" on 19 October 1987, when the market abruptly crashed, losing over 20 percent of its value and heralding a major U.S. recession. Even so, it was business as usual at Infinity's corporate headquarters, which meant more pressure on the local level. "Mel wouldn't accept any of those [excuses] . . . like business cycles, the 'Crash of '87,'" Berardini stated. Karmazin's enormous force of will, though, couldn't prevent the sluggish economy from pulling down profit margins at most stations and forcing a net loss at others. As Infinity's stock sagged, the chief remained adamant that the market was undervaluing his company, so with others in the corporate management team, Karmazin borrowed the capital needed to buy back Infinity, taking it private again in 1988. Berardini commented, "His attitude was, 'Fine . . . the market doesn't value us? We value us!' And he went out and bought all the stock back." Despite the increased debt load, Karmazin's strategy paid off as he waited out the recession, and his critics, and then took the company public again in 1992 for a huge cash windfall that he used, again, to pay down Infinity's debt and provide funds for expansion.

During the late eighties, WBCN continued to thrive, though the tough economic situation made the demands from corporate all that much more difficult to address. "[Tony] was the focal point for all that pressure." Bob Mendelsohn observed. "His whole job had become one of somehow squeezing all the parts together so that nothing would get hurt too badly and he could still give Mel what he wanted." Even so, Berardini's responsibilities increased dramatically in February 1987. After being promoted to a company vice presidency and retaining his general manager (GM) duties at WBCN, he was also given the responsibility of heading up KROQ in Los Angeles. Bob Mendelsohn smiled wryly at the memory: "Infinity was the land of the two full-time jobs. Tony did it. I did it . . . for a year and a half: general sales manager *and* national sales manager." The official WBCN press release stated that "Berardini will now direct the activities of both stations, dividing his time between Infinity's Boston and Los Angeles affiliates." This meant that the music geek from the Bay Area who had arrived at 'BCN in 1978 behind the wheel of a VW hippie bus was now going back out to the West Coast, wielding Infinity's hammer and sickle while still responsible for a station three time zones behind him. The bicoastal GM/VP started logging astronomical frequent-flier miles, jetting to LA for two weeks and then returning to Boston for the rest of the month. "Mel didn't tell me to do it that way; it was the pattern I established. I had a company car here and one out there [two 'heavy-metal' black Pathfinders]; I'd fly in on Monday, get there at noon—LA time, go right to the office and then work until six. Then the next day I'd be in the office at KROQ between 6:00 and 6:30 each morning; I'd deal with 'BCN issues and talk to them first, then KROQ's people would come in later."

Despite Berardini's physical absence half of the time and the world's economic woes, as WBCN reached its milestone twentieth birthday, morale couldn't have been higher. *Adweek* magazine ran a story that month entitled "WBCN Remains Solid as a Rock after 20 Years: Radio Advertisers Find the Boston Station a Sure Bet in a Fickle Market." The article cited 'BCN's remarkable ratings consistency: "First among men 18–34 and 18–49 for 22 Arbitron surveys running [four per year]. In the adult categories [men plus women] for those age groups, it has ranked first for 20 surveys." Governor Michael Dukakis and Mayor Ray Flynn both declared Tuesday, 15 March

Peter Wolf's Houseparty! Taking over the 'BCN airwaves in the name of rock and roll with "the platters that matter!" Photo by Mim Michelove.

1988, "WBCN Day" in the state of Massachusetts and the city of Boston, officially marking a farewell to the teens for "The Rock of Boston" and the opening of its third decade as an FM powerhouse. Peter Wolf returned to the airwaves for his annual "Houseparty" show of R & B hits and rock and roll memories, while Maxanne Sartori, Eddie Gorodetsky, and Danny Schechter also delivered guest spots throughout the day. The station's twenty-year milestone was underscored by a motor head's wet dream of a contest, the station giving away a 1988 Pontiac Trans Am as well as a sweet 1968 Pontiac GTO classic. In San Francisco, the annual Gavin Awards and Seminar for Media Professionals feted WBCN as national "Album Radio Station of the Year" for the third consecutive time.

Charles Laquidara, Ken Shelton, and Mark Parenteau had settled into every weekday for most of the decade; Tami Heide now handled nights beginning at six; Bradley Jay, at ten; and Albert O rode the a.m. hours from two until "The Big Mattress" crew once again tumbled in before dawn. "Weekend warriors" included former newsman Steven Strick (who had returned after four years to be a jock), Lisa Traxler (a ballsy midwestern admirer who had driven east unannounced to pitch Oedipus on a 'BCN gig), Shred (a dedicated local-music fanatic), Peter Choyce (known for his off-beat graveyard-shift witticisms), and "Metal Mike" Colucci (who produced Tony Berardini's "Heavy Metal from Hell" program). I became the station's full-time music director, replacing Bob Kranes, and trading my regular weeknight shift for a single one on Saturdays. The music meetings, fueled by our prevailing idealism to use this marvelous vehicle to turn people onto new sounds, continued. In Oedipus's case, he had a weakness for female artists (Tami Heide dubbed them "Oedi's Women in Pain"—and it stuck); Kathryn Lauren (who did nights for three years) loved any novelty record ("spastic plastic" is what she called them); Albert O might pitch a Robin Hitchcock single; Bradley, his eternal favorite Bowie; or I might be arguing for the new Big Country or the Alarm cut. Collectively, we actually liked it all (well, maybe not Kingdom Come—everyone hated them). "'BCN was fun and craziness, and a lot of political content too," Lisa Traxler remarked, "but music was an important part of the triad. Everybody, no matter what they might disagree on, had tremendous respect for the music, and musicians."

Tony Berardini, handling the business in his pressure-cooker general manager position, let off steam during his weekend "Heavy Metal from Hell" program. "Metallica came to the show in 1988 during the 'Monsters of Rock' tour; that was very cool. My favorite moment, though, was when Lemmy from Motorhead came up to be guest DJ. I figured he'd play all heavy stuff and Hawkwind, but all he wanted to spin was Motown!"

"Guns and Roses were in town playing the Paradise," Metal Mike remembered. Hired to the Listener Line in '86, he was filling in for Berardini on the air a few years later. "They had what they called a preshow press conference, which consisted of Axl and the guys pouring drinks in Stitches—the front room of the Paradise. I had just turned eighteen and Duff poured me a sixteen-ounce beer glass of vodka, then added just a touch of orange juice. This was three o'clock in the afternoon! Before I knew it, the 'Dise show was on; it was like a freight train running through the neighborhood."

"It was always the bands whose music I liked the least who turned out to be some of the nicest people ever," Tami Heide said. "Like Styx. They were so cool. Brad Delp from Boston, God rest his soul, he was the nicest person you'd ever want to meet. So I'd feel bad about not liking their music. But then I'd psychoanalyze it and think, 'Well, maybe their music *does* suck, and they have to overcompensate by being really, really nice!'"

Boston introduces *Third Stage* on WBCN in 1987. (From left to right) David Sikes (from Boston), Tami Heide, Kathryn Lauren, Brad Delp (of Boston), and Larry "Chachi" Loprete. Photo by Mim Michelove.

"I remember Brian Wilson when he came to the station," offered Albert O. "Brian was kind of fragile and had all of these handlers with him. He excused himself and went to the bathroom . . . and he didn't come out. The handlers were all pacing because they didn't have the balls to go in and interrupt him. He didn't come out for forty-five minutes!"

"I was supposed to do a backstage interview with Stevie Nicks," mentioned Steve Strick. "She walked in the room in a bathrobe with curlers in her hair, a little disoriented. 'Where are we now honey?' 'We're in Boston,' I told her. 'What station?' I showed her the WBCN emblem on the microphone and she said, 'Oh honey, I can't read that; you're gonna have to write it bigger.' So Bill Hurley, the Atlantic Records guy, took out a Sharpie and wrote 'WBCN BOSTON' in huge letters across the wall! That's how the interview began; then she just went off on Fleetwood Mac and how pissed off she was at Mick Fleetwood."

"Getting to meet Johnny Rotten Lydon was really cool," Heide offered. "We ate sushi with him and his wife. He swore it wasn't rice, but worms, in his sushi!"

"I had a great interview with Ziggy Marley, most of his brothers and sisters and Rita Marley," Ken Shelton mentioned reverently. "I'd seen Bob live at the Music Hall and I never got to meet him, but to meet his family! I felt like I was in the presence of the family of a god."

"Lisa and I kind of kidnapped Sammy Hagar and Eddie Van Halen one night," Heide recalled. Traxler confirmed the claim: "We gave them a ride back from their show at the 'BCN-trum' in a Subaru. It was a pretty wild ride on the Mass Pike; Michael Anthony passed us, driving by at eighty miles per hour, standing up in the sunroof of his white limousine and toasting us with Jack Daniels. It was a total rock star moment, a 'golden gods' sort of thing." She laughed at the memory and continued, "We decided we should do a call-in to the radio station, so we stopped at the Natick rest stop. Eddie walked up to the phones with me and I got Kathryn [Lauren] on the hotline. She immediately cut off the song that was on the air and we did some hijinks on the phone. But nothing happened that night. Suffice it to say, both Eddie and Sammy seemed surprised to have plenty of time on their hands after we dropped them at their hotel." Traxler chuckled, "We may have planted the seed for the song 'Finish What You Started.'"

Late in 1988 Peter Frampton had been booked for an interview with Parenteau, but the star was running late. When a listener called in to report that the guitarist was chirping happily away on the air at KISS-108, even though he was already supposed to be at 'BCN, Parenteau began berating the British star relentlessly on the air. Across town, Frampton finally finished his interview and hurried through the driving rain into his limousine for the twenty-minute ride to WBCN. Instantly the long car filled with a booming voice: "I don't care if he ever shows up at this point. It's obvious he considers the disco station in town more important than the rock and roll one." Frampton looked at his promotion man in disbelief. A couple of songs followed, but then the tirade resumed: "Let's call him disco Peter, if he ever comes at all!" By the time the limo arrived at WBCN, well over an hour behind schedule, Frampton had already decided not to enter the station. Parenteau continued to whip the affair up into a misadventure of dramatic proportions, goading the artist repeatedly: "Peter Frampton is finally here at BCN—well, he's actually outside in a limo and he won't come in!" While the Atlantic Records promo man tried to negotiate a truce and the musician stewed angrily in the car, no one from WBCN even saw him, except for sales secretary (and my future wife) Carrie Christodal, who slipped outside with Ed Maloney, a production assistant, to have their dog-eared copies of *Frampton Comes Alive* signed. There was no sense blaming these two; the star rolled the window down, smiled, and autographed the records.

A couple of weeks later, Frampton would protest his treatment in a personal letter to Oedipus, lamenting that a friendship with WBCN, which dated back to his Humble Pie days in 1969, had been trashed. Impressed with the star's candor, Oedipus resolved to bury the hatchet and invited Frampton back whenever possible. But it took five years before the planets aligned and Parenteau stood face-to-face with the guitarist. More than a foot shorter than the towering DJ, Frampton brazenly challenged: "Do you have something to say to me?" I hovered behind the two and prayed that Mark would take the high road. And he did, apologizing and, uh, eating humble pie. After that, all was forgotten as the pair became the best of "show-biz" buddies on the air. Frampton didn't even get mad when Parenteau busted his balls over the abominable *Sgt. Pepper's Lonely Hearts Club Band* film he did with the Bee-Gees.

Although WBCN had evolved greatly over the years, the station's involvement with musicians and artists remained consistent. The "Rock of Boston" concert showcased nine of the city's premier bands in a "Horizons for Youth" benefit at the Boston Garden on 2 December 1988. Long committed to promoting local music and, when possible, advancing careers to the extent of helping to secure record deals, WBCN celebrated each band's success story. Punk/R & B shouter Barrence Whitfield and the Savages opened the night, followed by New Man, national hit makers 'Til Tuesday, the Fools (with singer Mike Girard stripping right down to his underwear during the first song), O-Positive, Tribe, Del Fuegos, Extreme (on their way to a platinum future), and Farrenheit closing the show. The Cars' Greg Hawkes joined Tribe onstage, and singer Eddie Money took a moment before his show across town at the Orpheum to emcee for 'Til Tuesday. This success prompted a second "Rock of Boston" concert in October the following year. Once again a royal flush of A-list local bands, including Scruffy the Cat, performed, but the finale of the night beat the house as the Neighborhoods were joined onstage by the "Brads": Brad Whitford and Brad Delp, to rock out on a set of their signature Aerosmith and Boston songs, as well as a scorching version of "Communication Breakdown." The following November, the third annual "Rock of Boston" shifted focus from local talent to an international lineup, attracting ten thousand fans to see, not only the Boston groups Tribe and O-Positive, but also Irish bands the Hothouse Flowers and Something Happens, along with the Call and Masters of Reality.

There was a great deal of interest in Masters of Reality, a formerly un-known upstate New York band that Ginger Baker, the legendary English drummer of Cream, had joined. I was supposed to help out 'BCN's pro-duction of the show by hustling emcees on and off stage, but then Chachi walked up and told me there was no one available to pick Masters of Reality up at Logan Airport. I drove the Chevy Lumina (the same, lame, rear-wheel drive van that got stuck at all our ski events) down the Garden's long back ramp and over to the airport. After a short wait at the curb, the scruffy rock group tumbled out of the terminal and jammed noisily into the back of the van, Ginger Baker claiming the seat directly behind mine. While pulling out onto the access road, I could see the drummer's toothy wreck of a smile as he fixed my face in the rearview and said something completely unrecognizable in a mutated Cockney. I found myself wondering why this iconic musician hadn't taken some of the considerable royalties he must have made from *Wheels of Fire* back in '68 and repaired the few choppers left in his mouth. Perhaps Baker read my mind, because he started to mess with me. As we raced through the Sumner Tunnel, with the cement wall flashing past to the right, and cars whizzing by just to the left, the drummer reached forward and put his hands directly in front of my eyes! "Ahhhhh! What are you doing?! You trying to kill us?" I screamed.

"Ha, ha, ha," he guffawed in return, like some demented Sasquatch, and then yanked his gigantic paws away after a few seconds. In a morbidly perverse way, the only thing I could think about as we popped out of the tunnel and flashed past the North End to the Garden was how I could have made "Random Notes" in *Rolling Stone*. With my luck, though, the story would have been ironically titled "DJ Perishes as Rock Legend Survives Fiery Crash."

By 1991's "Rock of Boston" concert, 'BCN had completely filled the Bos-ton Garden with a bill that included Ray and Dave Davies of the Kinks, Foreigner, Brad Delp's new band RTZ, the Smithereens, and Boston's own Raindogs. The following year's Garden concert, number 5, featured the Spin

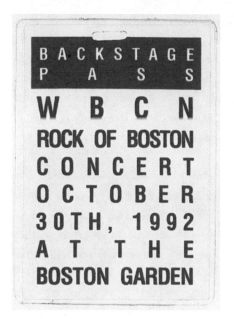

Backstage pass for the "Rock of Boston" concert, 1992. Phish packs the Garden. From the author's collection.

Doctors, Michael Penn, Material-issue, and WBCN's own foppish rock band of employees, the Stools (fronted by the Big Mattress's master writer and resident dry wit Lance Norris). Sales lagged behind expectations at first, but that situation abruptly changed when WBCN added a newcomer named Phish to the bill. The Vermont band's active network of fans had quietly gained such numbers and passion that, to everyone's amazement, the concert sold out within hours. Of course, that pattern would become typical; Phish filled up any place the band members chose to play during a long and acclaimed career. Jeffrey Gaines provided a highlight for WBCN listeners when Mark Parenteau surprised the singer as he walked offstage after his set and down the corridor to his dressing room. With wireless microphone in hand and live on the air, the DJ congratulated Gaines, who stopped on a stairway cradling an acoustic guitar, his shirt soaked through with sweat. Suddenly, the singer burst into song, crying out a rendition of "In Your Eyes." His version of the Peter Gabriel hit had become a runaway success story on WBCN, and the powerful delivery from the steps, echoing dramatically in the cinderblock-enclosed space, was captured by Parenteau, who held the microphone as steadily as possible till his arm turned blue and the hairs on everyone's arms stood at attention.

Live broadcasts of bands, a tradition since the earliest days, increased in frequency as the station presented classic performances from artists like John Hiatt, the Alarm, Bad Company, Lenny Kravitz, Little Feat, Iggy Pop, Sinead O'Connor, and Georgia Satellites. A broadcast series originating from Newbury Sound, across the street from WBCN, showcased local music and emerging national acts live on the air. By the end of the eighties, WBCN's engineering department had mastered the technology to pull off remote check-ins from just about anywhere in New England. When Great Woods, the outdoor arena in Mansfield, opened in 1986, the techies installed a permanent broadcast area backstage, just yards from the musicians' dressing rooms. The space also housed the venue's washer/dryer combination, which endlessly laundered sweaty stage outfits and soiled towels. If one listened closely, many of those backstage interviews included the sounds of a squeaky rotating drum and thumps of heavy clothes or the occasional pair of sneakers in the background.

The Great Woods' local crew and some bands themselves often utilized the laundry room/broadcast studio, located just a few yards behind the stage. This led to some surprises, as when Van Halen marched into Mansfield for a show in 1993. When the 'BCN technical crew arrived, they discovered that their broadcast area had been commandeered and transformed into a huge pyrotechnics supply depot. "No, no, no! You absolutely cannot be in there," the band's stage manager yelled. Protests were lodged with Tom Bates, Don Law's captain on the ground, and WBCN struck a deal with the reluctant visitors. Only under the watchful eyes of Van Halen's crew, who would have shared the ride to the hereafter if a mishap occurred, was the 'BCN team permitted to check in live, amidst a potentially lethal amount of flash pot ammunition and explosive charges piled high in boxes all around them. Needless to say, neither Sammy Hagar nor Eddie Van Halen could be enticed to do an interview from the powder keg.

It didn't get any bigger than Foxboro Stadium, the home of the New England Patriots, and WBCN broadcast live from the venue on Aerosmith's 1986 "Back in the Saddle" tour; for U2 the following year; from 1988's "Monsters of Rock" show; for a pair of Who reunion shows in '89; on three wintery nights that same year with the Rolling Stones on the "Steel Wheels" tour; and at Paul McCartney's return to the stage in 1990. John Mullaney, hired as a broadcast specialist for dozens of remotes, mentioned, "At Foxboro, we put a transmitter in a golf cart and sent the signal up to the press box where we placed the equipment to hook into the phone lines back to the city." No longer tethered exclusively to backstage phone lines, WBCN's jocks now roved freely through the parking lots at Great Woods and Foxboro in full view of its tailgating audience. As soon as concertgoers realized that the vehicle closing in on them did not contain security personnel or police but instead sported a WBCN logo and

Broadcasting live from "Monsters of Rock," Foxboro Stadium, 1988. The WBCN remote crew hangs with Van Halen. Notice how each member of the band has to have a radio lady next to them (it's in the rider). Courtesy of WBCN.

conspicuous antennae array, a sea of people would surge forth to greet the radio envoys. "Once the people got in front of you, you were kind of done, 'cause you couldn't run them over!" Mullaney laughed. Like a political candidate wading into a forest of campaign supporters, they found themselves completely inundated. "Hey 'BCN! Can we go on the aaaaair?" "Want a hamburger?" Broadcast producer Jefferson Ryder remembered, "Golf carts were never designed for this; three radio guys, two hours, and one objective: get real fan interviews and get out alive!" Ryder dubbed one particular section of the Foxboro parking lots "The Gauntlet," because of its triangular shape and difficult traverse. The cars parked there had to be arranged in rows that rapidly pinched closer and closer, constraining forward progress. "You couldn't go zooming through there, so people would just crush in around you," Mullaney added. "Plus I think they parked those folks in there first, so they were a lot more inebriated than the rest of the audience!"

Although none of the merrymakers at WBCN knew it, this was their last chance to drink up, metaphorically speaking, because other partygoers were about to crash the open bar. The first significant and dangerous challenge came from WZLX-FM, a station that signed on the air in October 1985 with a brand-new format playing "Classic Hits." Their rationale was that a large number of listeners had grown up on the rock songs of the sixties and seventies, and were unhappy with the recent MTV-spawned Top 40 dance fare, like Duran Duran, or the blight of "hair band" metal. This proved to be an inspired guess, and WZLX experienced immediate acceptance and rapid growth. *Billboard* magazine reported that in January through March 1986, which would have been the first complete ratings period with the new format, the station had blasted off from a number of 3.1. to an impressive 5.0 share in all listeners over the age of twelve. In men eighteen years and older, WZLX raised eyebrows by jumping from a 3.9 share to a striking 6.9. In only five months, the groundbreaking "Classic Hits" station stood second in rank to WBCN in the important "men 25–54" demo. *Billboard* continued, "WZLX program director Bill Smith says his station's Arbitron gains have been achieved at WBCN's expense. 'We're delighted with the results,' he says. 'We have so many people calling the station who admit they switched from WBCN.'" At first, there was little reaction at "The Rock of Boston," although Oedipus did experiment with the term "vintage" in labeling 'BCN's own heritage music selections. That didn't resonate with the station's listeners, though, so the imaging term was soon dropped.

To general sales manager Bob Mendelsohn, the arrival of WZLX meant a daily competition for ad dollars. "I was very concerned when Oedipus made the conscious decision that WZLX's presence wasn't going to change anything. It's true that in the mideighties we had true market dominance, which would take care of most ills, but we maintained that same posture when suddenly there was a main competitor that was hungry for the business." As a radio professional hired to WBCN by Mel Karmazin himself, Mendelsohn already had problems with the way the station was run: "When it came to the business there, anybody in a senior position thought that just because the station was cool, people should be flocking to it—which was a serious and fundamental lack of insight. Any client doesn't really care about how cool the company is. That may get their attention, but then you have to perform the service for them that they paid you a lot of money to do. This is one of the reasons that Oedipus and I, particularly, never saw eye to eye on anything."

Another challenger emerged in the later eighties to joust with WBCN. Located fifty miles west in Worcester with a signal that barely penetrated into downtown Boston, WAAF-FM competed aggressively for the large Metro West population living between the two cities. With a mainstream approach not unlike WCOZ from just a few years earlier, 'AAF played harder-edged album-oriented rock (AOR), avoiding the pop music sounds by artists whom WBCN regularly embraced (and even helped to break), like Cindy Lauper, Icehouse, the Bangles, or Simple Minds. However, 'AAF's unrelenting diet of Van Halen, AC/DC, Loverboy, and so forth, achieved a focus that WBCN, for better or for worse, comparatively lacked, and the station's approach began to bear fruit. In the winter 1986 ratings for Boston, the station found favor with a younger male audience, slipping under 'BCN's radar to actually claim first place in "men 18–24 years old." True, the station appealed to a less affluent, less desirable demographic, but its managers could now claim a sizable gain in a significant niche by concentrating on a narrow selection of elementary "meat and potatoes" rock. *Billboard* reported in May 1986, "According to WAAF program director Cynde Slater, the Katz Broadcasting facility 'made a commitment to go after the Boston market two years ago.' She says the only competitor the station targets is WBCN." During these early days of a rivalry that would extend for over twenty years, 'AAF showed a "whatever it takes" mentality and single-minded concentration on undermining the 'BCN brand. *Billboard* continued, "At WAAF promotions are aimed directly at the WBCN listener. A recent television commercial included the following dialog: 'BCN's a great station; I just want to rock. Enough talk!' Later in the spot, a WBCN bumper sticker is covered over by a WAAF bumper sticker." Although WZLX represented the much more immediate and dangerous threat, certainly to 'BCN's older listeners, WAAF's attack on the younger audience would be impossible to ignore.

Then, a third opponent emerged, a radio station far less menacing than either 'ZLX or 'AAF, but you'd never know it by the amount of press that the place received. "WFNX-FM, based in Lynn, is the new kid on the block, a David to WBCN's Goliath," wrote Jim Sullivan in the *Boston Globe* in January '86. "Formerly WLYN, the station was bought by *Boston Phoenix* publisher Stephen Mindich in 1983. WFNX is a 3,000-watt station that has made a small dent in the Boston market, but arguably, a larger dent in its mindset."

Sullivan pointed out that WFNX's ratings were less than a ninth of WBCN's but postulated that the Lynn station's choice of music carried a far bigger stick. Playing edgy new sounds in a genre that would eventually become known as "alternative," WFNX eroded 'BCN's reputation as the station to turn to for the latest music . . . at least to the people who cared little for mainstream and who could actually pull in WFNX's limited signal. The community of Boston's writers and critics, inclined to praise advances in the arts and solidly ambivalent to the status quo, lavished attention on their new "David." So, while WFNX could not challenge 'BCN on a business level, it certainly added torment to a station now locked into protecting its other flanks against more commercial competitors.

The WBCN slogan, "Classic to Cutting Edge," aptly described the station's musical mission since 1968, but it also neatly defined the flanks of the radio war raging twenty years later. On the right, WZLX played classic AOR fare, while WAAF concentrated on their basic thud rock; plus, WODS-FM featured a sugary-sweet mix of pop oldies delivered by a team of ever-cheerful, over-the-top, hyperactive announcers. These stations kept up the mainstream pressure while WFNX fought for at least an ideological left. Was it possible for 'BCN to remain standing on its original base of the Beatles, the Stones, and The Who while covering the younger trailer park, roadhouse styles of Lynyrd Skynyrd, AC/DC, Def Leppard, and Ted Nugent? And if succeeding at that, how could 'BCN include the leading alternative edge with bands like R.E.M., the Cure, Red Hot Chili Peppers, and XTC (many of which 'BCN broke in the States)?

In 1989, Jeff McLaughlin assembled an article in the *Boston Globe* about the ever-increasing level of competition and changing tastes in the market. He wrote, "WBCN is 'The Rock of Boston,' positioning itself as the station that defined rock culture in the metropolitan area. But as its core audience grew older, it became stretched thin, its position harder to hold. The quirky pop favored by today's 18-year-olds doesn't find much favor with 40-year-olds in business suits, and, conversely, young people interested in the cutting edge of pop music can only tolerate a small amount of music from what they call 'the dinosaur era' of rock history." Although the station's famous assortment of veteran personalities and a strong promotional presence had been compelling enough to retain an impressive audience within WBCN's orbit for years, now the strategy of its music mission was in question. Could the station continue on its all-encompassing course and maintain the ratings success to which it had become accustomed? John Gehron, the general manager of WODS, didn't think so. In the *Boston Globe* article, he told McLaughlin, "You can take two approaches. You can grow older with your audience, or let the succeeding generational waves flow through you. You can't do both."

But WBCN tried, toeing the mainstream line while embracing the "cutting edge." In 1987, Oedipus established "Boston Emissions," a Sunday-night radio show devoted wholly to local music, as a means to further concentrate on the Boston music scene. Following his beloved "Nocturnal Emissions" program, during which Oedipus featured the alternative music he had always been most fond of, the new local music show furthered 'BCN's commitment begun so many years before in the Rock 'n' Roll Rumble. "Everywhere in Boston, people were dressed in black and carrying guitars," Lisa Traxler commented. The interesting thing was that even though WBCN's bread and butter came from the big ticket mainstream bands, its DJs were mostly interested in the music emerging from the street and college radio, not surprising since most of them came from there. During this period, the air staff prided itself in helping to establishing careers for new groups like the Alarm, Smithereens, Living Colour, Black Crowes, Georgia Satellites, and the Cult, while also comfortably programming Don Henley, Stevie Nicks, and Pink Floyd. *Virtually Alternative* reported in its thirty-year station retrospective in '98, "WBCN was still on the edge when compared to the national AORs, championing bands like the Godfathers and An Emotional Fish to go with Led Zeppelin and Aerosmith. Their bread and butter, however, was the same classic rock that fed the rest of the AOR world. Beneath the surface, the new musical movement was brewing. When it finally arrived in 1991, it was championed not by the historically progressive WBCN, but by the little station on the North Shore with its collegiate attitude and small signal—WFNX."

"Oedipus always did such a great job taking what we considered to be our big artists out from under us," mentioned Bill Abbate, who started at WFNX in 1984 and jumped ship to 'BCN four years later. "A great example of that was in '88; [at 'FNX] we felt that we had done everything we could in the marketplace to champion the Smithereens. But we didn't get the big show. 'BCN got them instead for their twentieth-anniversary, monthlong celebration." Oedipus paraded 'BCN's clout and transmitter signal in front of the record companies and nearly always received cooperation. Abbate continued, "There's no doubt that 'BCN played [the Smithereens], but the feeling in Lynn was, 'Hey, we've got them in full rotation; this is our band!'"

When another new group named Nirvana, and its anthem "Smells Like Teen Spirit," hit the streets, a new movement arrived at the gates with a battering ram. Rarely can a shift in taste and a changing of the guard be measured in such a distinctive and singular event, but here it was; a new generation had advanced onto the field and with this song made its presence known. Mark Kates, head of alternative promotion for DGC, the record label that released Nirvana's *Nevermind* album, remembered that WFNX and its program director Kurt St. Thomas had a relationship with the band well in advance of the album's appearance: "The band was deeply entrenched at the station; they would have played 'Teen Spirit' the moment it came out. The album was released on September 24, 1991, so the single came out in August." Nirvana had first played the Jamaica Plain rock club Green Street Station in '89, then Cambridge's ManRay the year after, a period in which the 'FNX program director became a devoted fan and acquaintance—so much so that Thomas prevailed on the group to appear at his station's eighth anniversary party the day before *Nevermind* even hit the streets.

WBCN would not ignore "Smells Like Teen Spirit," playing it in October after the single's video had hit MTV's "Buzz Bin" and well before most mainstream stations. However, to many in the local scene, this was seen as a game changer, WFNX being the point station at this crucial cultural juncture in the same way that 'BCN had been so many times before. "Back then, there was a sense of purity that if a song was so good, there was no way it wouldn't make it to the air. 'Teen Spirit' was that to me, a no-brainer, and we were late to the party," David Bieber admitted. "Nirvana actually came to the station and they weren't even put on the air. They were given a back-alley treatment and cut some station IDs." But to be fair, Mark Kates brought Nirvana by WBCN before DGC had begun actively promoting the group, simply because the band happened to be in Boston. This was a courtesy call for Kates to put in the bank for when he'd later be looking for support from Oedipus. The band members dropped by 1265 Boylston Street on a Sunday, when few 'BCN staffers were present. With Nirvana barely known and their record not on the air yet, the group was not invited to do an interview. In similar situations, the station would not have been accused of missing anything; however, after the band's enormous and immediate impact, WBCN was perceived as being behind the ball in Boston. As Nirvana's presence only grew and launched a tsunami straight through the heart of rock music, the band member's appreciation for the earliest stations that had championed them would never waver. Kurt Cobain remained a friend of 'FNX, not WBCN, till the day he died.

NELSON, HOWARD, AND "THE LOVE SHACK"

After years of gazing out over a lunar-like no-man's-land walled by concrete and barbed wire, the East Berliners who crowded against the barrier erected by the Soviets in 1961 could never have imagined the magnitude of lights, sound, production, or the simple freedom that Pink Floyd's Roger Waters enjoyed, to stage his band's grand 1979 work, *The Wall*, in Potzdamer Platz on 21 July 1990. Only the previous week, West German troops had finished sweeping the area where two hundred thousand fans now stood for unexploded bombs and mines. Their Eastern counterparts danced atop a long line of armored personnel carriers, hooting and hollering in their fatigues as Waters unwrapped the iconic double album onstage, and East hugged West in a new summer of love. I journeyed to this massive European party as a WBCN correspondent, phoning reports back to America from a bank of phones set up in a trailer jammed against the Berlin Wall. After nearly forty years of Soviet injustice and Hitler's nightmarish reign before that, this whole area reeked tragically of history, like some decaying roadkill left by a continent-wide collision. But the immense sound of nearly a quarter-million voices booming into the Berlin night: "Tear down the wall! Tear down the wall!" shouted down the past. Could it be? Could a glimmer of hope be shed for these Berliners, that the end of their long nightmare was truly at hand?

Afterward, a few of us roamed about in a mostly deserted backstage area. Dave Loncao, a high-level Mercury Records rep, flashed a laminated pass and performed his Big Apple razzle-dazzle to secure a few bottles of red wine, all that was left from the considerable bar that had catered earlier to hundreds of VIPs. We drank enthusiastically, talking about the concert and the politics of a new Europe until the last bottle had been drained. Loncao grumbled, "What do you want to do now? We'll never get out of here with all the traffic!"

"I guess we just have to hang out or start walking back to the hotel," someone offered, not a happy prospect since the place was several miles away.

"Maybe we should just walk over and take a piss on the Berlin Wall!" someone else piped in. I'd like to say it was my idea, but the wine had begun interfering with any accurate record keeping by that point. Let it stand that "someone" in our group had the brilliant thought.

"Like the cover of *Who's Next*?" Loncao observed, chuckling with a hand already on his fly.

"Exactly!" And with that, our little band of American broadcasters lurched over, offering up our own special tribute to the absurdity of building that ugly ribbon of concrete and metal in the first place.

CARTER ALAN

A new decade had arrived, a noisy and kicking brat named 1990. Loud and brash, the imp quickly drowned out its older, more reserved, brothers and sisters born during the eighties. Just the end of the Cold War, the breakup of the Soviet Union, and the reuniting of East and West Germany was astounding enough to warrant special attention. But to that remarkable turn of events, add Lech Welesa's presidential victory to break the back of Poland's communist government and Margaret Thatcher's resignation. Twenty-five cents bought a first-class stamp and a gallon of regular gas averaged around $1.15. Sporting fans watched Edmonton deny Boston the Stanley Cup (again); the Reds swept Oakland in the World Series; and Joe Montana led the 49ers in a rout of Denver at Super Bowl XXIV. *The Simpsons* debuted on Fox-TV, *Seinfeld* on NBC, while *Home Alone*, *Dances with*

Big hair's last stand: the eighties give way to the nineties. Carter Alan and (future) wife Carrie Christodal. Photo by Roger Gordy.

Wolves, and *Ghost* became box office monsters. As M. C. Hammer's *Please Hammer Don't Hurt 'Em* sat at number 1 for twenty-one weeks and sold ten million copies, an unknown band in Seattle named Pearl Jam played its first gig and Nirvana visited Boston for the second time.

The eighties had been very good to WBCN, the station arriving at 1990 on the crest of local popularity, as well as national respectability. *Friday Morning Quarterback* looked back on the previous years by polling the professional communities of both broadcasting and the record business nationwide, naming 'BCN as "Best Station of the Decade." In the 1990 *Rolling Stone* Reader's Poll, and for the ninth time in ten years, the station was voted one of America's best large-market radio stations. But just as the fresh year had brought sweeping changes to world affairs, politics, and culture, 1990 also perched 'BCN on the edge of a worrisome and uncertain future after a decade of consistency and mastery of its domain. Several personalities departed from the air staff, including Tami Heide, who transferred to KROQ for a long and acclaimed career, and Lisa Traxler, who ended up on rival

WZLX in 1992. Billy West and Tom Sandman both exited the production department, the latter stepping up to a program directorship at crosstown WBOS-FM: "I didn't see any future in me getting into programming at WBCN; I didn't think Oedipus was going to go anywhere soon." Sandman was correct; 'BCN's now-veteran program director remained secure in his position, presiding over the latest changes in his staff, which would prove to be minor compared to the seismic ones arriving in just a few years. Tony Berardini endured as station general manager and Infinity vice president but, by the end of 1989, had hired a general manager for KROQ and given up the arduous schedule of flying back and forth to LA every two weeks. "The real issue became that 'BCN started getting more competition from 'ZLX," Berardini explained. "That station was chipping off enough of our ratings that it made it difficult to reach the budgets. Now I could sit down and focus on 'BCN."

In the news and public affairs departments, the twin defections of Katy Abel and Matt Schaffer left a gaping hole. Oedipus prevailed upon Sherman Whitman, who had been part of the WBCN news team for three years until October 1987 when he left to work at WXRK in New York City, to return as news director. Whitman found the new appointment satisfying: "There was still a commitment to the news and public affairs at 'BCN; people felt they got information there that they couldn't get anywhere else." Then, Oedipus tapped Maurice Lewis, who sported an extensive list of credentials in local radio and television journalism, to take over for Schaffer on the "Boston Sunday Review" (BSR). Lewis said, "I had total and free reign without any interference from management; that was the beauty of the show. You couldn't have asked for a more supportive environment." The two new arrivals, along with Charles Laquidara, soon brought great honor to WBCN in June 1990, on the occasion of black political activist Nelson Mandela's historic visit to Boston. Sadly, as triumphant as that moment would be, it would also represent the final hurrah for an acclaimed (and some would say, essential) part of WBCN's original manifesto in 1968.

After twenty-seven years of harsh imprisonment at the hands of the pro-apartheid South African government, Mandela had been released from prison as internal hostilities and international pressure mounted on the white ruling party to hand over power. Eventually, a multiracial government would be installed with Mandela as its head, but in the immediate afterglow of his newfound freedom, the most famous and revered symbol of racial equality since Martin Luther King Jr. decided to embark on a goodwill tour of America. The important and historic meetings with national leaders and fellow activists would mostly occur in New York City and Washington, D.C., but Mandela planned a one-day, whirlwind visit to Boston, which generated extraordinary excitement from ordinary citizens and local politicians alike. Whitman enthused, "Katy [Abel] was gone by then, but she had done "Commercial Free for a Free South Africa" in 1985, and to think that Mandela himself would come to Boston five years later!"

As the time wound down swiftly to the historic visit, and on the day before the African leader's 23 June arrival, WBCN was recognized for its role in popularizing the anti-apartheid movement. The Boston City Council surprised Charles Laquidara with an award for "his Commitment to Ending the Apartheid Regime in South Africa," as a result of the on-air campaign to boycott Shell Oil. This was followed by a second citation from the Massachusetts House of Representatives in acknowledgment of his "Continuous Public Education and Support for a Free South Africa." A similar award went to Maurice Lewis, as representative of the public affairs department, in recognition of WBCN's efforts toward these initiatives. While the accolades were being handed out and photos snapped, Sherman Whitman worked feverishly down the hall, making final preparations with the station engineers for WBCN's in-depth coverage of Mandela and his wife Winnie's visit.

Maurice Lewis had helped to coordinate a simultaneous playing of the South African national anthem on various Boston radio and television stations at 12:01 on the day Mandela arrived. "It was the first time we got competing stations to go along with each other to do anything!" Lewis laughed. "Also, working with [Urban radio station] WILD, we were involved in organizing a parade launched from Roxbury, which wound its way through Back Bay and over to the Hatch Shell." After participating in the motorcade, Lewis arrived at the massive tribute concert on the Esplanade where over three hundred thousand adoring citizens waited, part of a visit described by *Ebony* magazine as "Mandelamania." "It was a Saturday afternoon when Nelson Mandela came and spoke at the Hatch Shell," Whitman recalled. "We were there from the start of the day to the finish. We carried it *all* live: Mandela's entire speech, the music the artists were playing, and the words of Governor Michael Dukakis and Senator Ted Kennedy. There was a sea of Mandela T-shirts out there . . . like seeing a rock star." Meanwhile, Lewis, astonished at the size of the crowd that lined both sides of the Charles down to Massachusetts Avenue, appeared several times on the big stage. "I was called back multiple times because the Mandela entourage was late. I received those dreaded instructions: 'Fill! Fill!'" he laughed. "But because of that I got the chance to 'fill' onstage with Danny Glover and Harry Belafonte; those [moments] were great."

This day, auspicious as it was, would end up being the last big hurrah for WBCN news, signaling the twilight years for the department as 1991 arrived. First, Lewis decided to move on, with no regrets: "It felt great to work on a show that brought opinions, cultures, theater, drama, politics, and diverse people from Boston together; that's what the 'Boston Sunday Review' had always been designed to do." During the same period, as Whitman described, his role at the station changed: "[Oedipus and Tony] said, 'We're going to have you stop doing the news and let Patrick Murray handle things." Murray, who had started off two years earlier as a news department trainee and Whitman's intern, was surprised by the move. He mentioned, "I didn't know about it beforehand; Oedi just called me into his office and told me, 'You're the news director.'" Murray had worked his way up to be a well-known morning sidekick on Laquidara's show, due to his offbeat and free-spirited delivery of news, traffic, ski, and surf reports. "It was a time when you could get your news in other places, so on the Mattress we did it differently. I'd make sure the report was on the pulse of what was going on and then have some fun with it, like a *Saturday Night Live* sketch." Though the official title of news director had been conferred upon him, Murray didn't necessarily see much increase in the scope of his responsibilities, nor a directive from Oedipus to change his style. He continued to gather information for Laquidara and deliver it with his wise-cracking, sardonic slant. "Once, as I was getting ready, I noticed they had the agricultural report up on the AP newswire. So, I did hog slaughters, sheep slaughters, the rye crop, and the wheat crop numbers instead of a business report. We played a little *Raising Arizona* bit behind it with some sound effects. No biz report, just the cattle kill for the week!"

As Murray churned out his comedic, and more digestible, version of "the news," Oedipus unveiled his plans for the displaced Whitman, who explained, "With Maurice gone, they needed someone to do the BSR, so I handled that for the rest of my tenure there [till June '93]." Whitman did not turn bitter over these changes; he realized that the times were changing and that he personally couldn't stem the tide: "Every music station had begun cutting back on its news commitment, especially on the FM dial. There was specialization: music stations would now be doing music, and news stations would now do just news; 'BCN was no exception to that." In an ongoing series of deregulation moves in Washington, the FCC granted radio station owners far more latitude to run their businesses as they saw fit. The relaxed state of ownership rules and public affairs requirements set off a wave of layoffs as companies raced to reduce their bottom lines by slashing staffs, consolidating stations, and standardizing formats. By having Murray handle "Big Mattress" duties and also the news, WBCN management eliminated a full-time job and maintained spending on its Sunday newsmagazine show at a part-time level. "I turned out the light in the WBCN news department," Whitman acknowledged. "And that was one of the saddest things to do, simply because, for me, 'BCN was home [and] all the people there . . . that was my family. All the things we did there together, at 1265 Boylston Street, we did to make radio great. Somewhere down the road, someone made the decision that we don't want to be great anymore; all they cared about was making fabulous profits. I left because I felt my time had come to an end." Not to take anything away from Murray, who was given the football, ran with it, and scored points, but from this point on, news "dissecting" on WBCN was a thing of the past.

On 28 February 1993, WBCN's staff gathered at the Hard Rock Café in Back Bay to celebrate the station's twenty-fifth birthday. Invitations had gone out to current workers as well as all the former employees who could be found. "It began with Carla Wallace, an original 'BCN person at Stuart Street," David Bieber explained. "She called me in December '92 and said that we should do a twenty-fifth-anniversary party." It seemed like an obvious idea, but Wallace had the additional vision that the party should be open to anyone who had ever worked for 'BCN. The happy result was that four hundred employees jammed into the Hard Rock, their buzzing conversations easily drowning out the blaring music, that is, until the members of the J. Geils Band found their way to the stage and serenaded the crowd with an a cappella version of "Looking for a Love." It was the band's first public performance since it had broken up a decade earlier and a symbol of respect, not only for singer Peter Wolf's former radio home, but also for the long lineage of legendary characters that had inhabited 104.1 FM. Stars from former times arrived in force: J.J. Jackson, Tommy Hadges

T. Mitchell Hastings addresses the crowd at the Hard Rock Café in 1993 for WBCN's twenty-fifth anniversary. Photo by Dan Beach.

(now president of Pollack Media Group), Billy West (the voice of the *Ren and Stimpy* show), WBCN's founder and original owner Mitch Hastings, Danny Schechter, John Brodey (a general manager at Giant Records in LA), Norm Winer (a program director in Chicago), Maxanne Sartori, Ron Della Chiesa, and Eric Jackson. Aimee Mann, James Montgomery, and Barry Goudreau of Boston mingled about. Susan Bickelhaupt queried in her *Boston Globe* article about the anniversary, "25 years later, what has replaced rock station WBCN-FM? Nothing. It's still 'BCN, 104.1. Sure the jocks are older and drive nicer cars, and you can even buy stock in parent company Infinity Broadcasting, which is traded on the NASDAQ market. But the station that signed on as album rock still plays album rock, and is perennially one of the top-ranking stations and a top biller."

A rumor passed about the room that night, given weight since it allegedly came from a WBCN manager: something major was going to happen with one of the station's primary weekday air shifts. Since lineup changes at the station occurred with the frequency of Halley's Comet appearances, this warranted front-page news (or, at least, endless speculation). Was a member of the famed daily triumvirate of Laquidara, Shelton, and Parenteau finally bowing out? Oedipus and Tony Berardini weren't talking, but their conspicuous silence seemed to support that something big was about to happen. Industry analysts caught a whiff of the rumor as well, but it came from a different source: New York, on the nationally syndicated Howard Stern show. In the years since the former Boston University student had attempted to find work at 'BCN, the announcer had made his mark outside of Boston, rapidly rising to become a hurricane force in radio with his undeniable wit, lapses of taste, and incessant sparring with the FCC over established indecency boundaries. The shock-jock had always desired to return to Boston, hinting that this might be a possibility in the near future, but the official confirmation of that did not arrive until 8 March 1993. That night, WBCN began airing Stern's nationally syndicated morning show, on tape delay, following Mark Parenteau's afternoon shift.

Tony Berardini told *Virtually Alternative* in 1998, "Howard Stern was doing extremely well at Infinity's WXRK, and then at 'YSP, Philly. Oedipus and I kept looking at it, thinking, 'Wow, the company is going to want to syndicate him in Boston, too, why wouldn't they?' It was one of those things we knew we were going to have to deal with, yet Charles Laquidara still had good numbers in the morning. Even Howard used to listen to him when he went to Boston University. With this in mind, we went to Mel with the idea of running his show by tape delay at night. We had an immediate opportunity there, and we decided to take advantage of it."

"Howard wanted to be on the radio in Boston, and I was able to make that happen because I could put him on at night," mentioned Oedipus. "Everywhere else [with the exception of KOME in San Jose], he did the mornings, but Mel allowed me to make the decision to tape the show and play it later." Oedipus did not regret his decision: "Howard is a monumental talent; it just can't be denied. I remember many a time driving home from work and then just sitting in my garage, not getting out of my car, because I didn't want to miss a moment." Listeners tended to react to Stern's show simply and starkly: they either loved the DJ or hated him. Those who disliked his presence on WBCN found other nighttime options, but those attracted to his brand of radio comedy became riveted . . . for long periods of time. "Nighttime listening to rock stations was really falling off at that time, and we had nothing to lose," Tony Berardini told *Virtually Alternative*. Very quickly, as nighttime ratings at WBCN increased dramatically, the gambit, it seemed, had played out to be a shrewd and successful decision.

"Metal Mike" Colucci was one of the part-time employees now called upon to man the studio each weeknight, acting as a board operator during the taped Howard Stern program. Although just a lowly, anonymous worker bee to the mighty Stern, Colucci would be elevated by that jock's rabid following to the status of infamy. This was one of the secrets of that show's success: the shock-jock's ability to whip up drama, creating form and substance out of something that, when examined more closely, really warranted no such time or attention. Who cared? In Stern's hands, though, it seemed that a lot of people actually did (at least for four or so hours a day). On the evening of 27 July 1993, Reggie Lewis, a twenty-seven-year Boston Celtics rising star, collapsed on the court and died from a heart defect while Howard Stern's taped program ran on WBCN. Shocked, Mike Colucci phoned Oedipus to ask what to do. "He told me, 'Fade out Howard, play "Funeral for a Friend" [by Elton John] and announce that Reggie Lewis is dead.'" However, as an inexperienced DJ, Colucci couldn't handle the wave of emotion that swept over him as he spoke on the air: "I got all choked up; I went into this terrible tailspin before I got back into Stern's show as fast as I could." Unbeknownst to the young board op, one or more of Howard Stern's ardent Boston fans taped the show and overnighted a recording to the shock-jock for his live broadcast the following morning. Stern played the tape on the air relentlessly, downplaying the drama of Lewis's death to focus on Colucci's painful on-air gaffe. "It turned out that he did forty-five minutes on me, saying things like, 'If this guy is acting, he's a genius.'" Red faced, he endured the episode over and over again, taking a lot of good-natured ribbing from a host of people who never even heard the original episode. "At the time, Howard had the biggest radio antennae going; nobody was going to break in on his broadcast. He could stick it wherever he wanted."

Many employees and fans of the station disagreed with the strategy of adding Stern to the WBCN lineup, finding the move troubling, at the very least. Bob Mendelsohn, as general sales manager, would work intimately with the new syndicated show, selling airtime to local and national clients: "The first time I heard Howard Stern, I hated it. It just embarrassed me. It wasn't his sense of humor; it was his complete elimination of standards. These were the boom years; everything was about stock price and profits now."

"When Howard Stern got [to 'BCN], it was all about how much money he was making [for the company]," Danny Schechter observed. "These were the values of Mel Karmazin. They didn't really care about Boston, [and] they didn't really care about the audience; it was the market and getting market share. That's what mattered to them: it was about the shareholders."

"I know a lot of people mark Howard Stern as the beginning of the end, and maybe it was," Tim Montgomery said. "Because, in a way, Ken Shelton and his Lunchtime Salutes [and] the crossovers between Charles, Ken, and Parenteau: that represented the old 'BCN. God, there was some brilliant radio! Then, with Howard Stern and the whole coarsening of the culture, the station took on that mean, macho, sexist [attitude] with eighteen-year-old-boy fart humor and vagina jokes. That's not what 'BCN was! It was smart; then it became stupid. The dumbing down of the 'BCN audience became the whole new thing."

"That Mel had this amazing asset in New York and then was able to multiply that asset by putting it on all these stations, as a corporate move, bordered on brilliant," Bob Mendelsohn admitted. "But it was *everything* that 'BCN had *never* been before. WBCN had always been credible and sincere; this was smart-ass radio. As an employee of the station, that's when my affection and my pride in working there started to go in a different direction."

A second seismic event that shook WBCN in 1993 had been simmering since the year before, when Infinity Broadcasting, which owned and operated eighteen radio stations at this point, had entered into an agreement to purchase WZLX-FM in Boston. The $100 million deal with Cook Inlet Radio included the acquisition of two other radio properties in Chicago and Atlanta, and was made possible by another relaxed ownership edict from the FCC that increased the number of radio stations a company was permitted to own. Previously, a single operator could only possess a total of twelve AM stations and twelve FMs, but those limits were increased to a total of eighteen AMs and eighteen FMs. Additionally, the previous rule that a single operator could own a maximum of one AM station and one FM in a single market was increased to two each, as long as there were fifteen or more radio properties in the area and that listener share did not exceed 25 percent of the market's total. Infinity chomped at the bit to take advantage of these revised, business-friendly rules: although the FCC edict did not go into effect until 16 September 1992, the company proudly trumpeted its pending purchase in an 17 August press release. Obtaining final FCC approval would push Infinity's official takeover of WZLX into '93, but when all was said and done, WBCN's biggest rival had suddenly been made a member of the same family. The two stations had been kicking each other under the table since the mideighties, but Mel Karmazin now demanded that the horseplay stop immediately. "The day that [Infinity] closed on buying WZLX, [Karmazin] got the sales managers from both stations together for a meeting," Mendelsohn recalled. "His message was, 'You guys have been out on the street trashing each other for years. As of today, that stops; you're working together.'"

Not only did the new association with WZLX mean that the sales strategies at both stations could be allied and results maximized, but Oedipus now had the opportunity to collaborate with his former adversary, program director Buzz Knight, in working out a two-station approach that minimized the considerable musical overlap of their respective formats. This appeared to be the answer to WBCN's programming dilemma of addressing the tastes of two steadily diverging radio audiences. "This is what happens when you have a radio station whose life is so prolonged; you outgrow your audience," Oedipus pointed out. WBCN could either grow old with those longtime (twenty-five- to fifty-four-year-old] fans, becoming essentially a classic rock radio station, or embrace the younger listeners who were moving into the station's demographic target. With WZLX in the mix, the decision essentially became moot. "We elected to stay younger; WBCN had to. We focused on the eighteen- to thirty-four-year-olds." Making that transition would occupy a year or so, but the first adjustment, proposed almost immediately, became the station's third seismic shakeup in 1993.

Along with conducting the necessary music and image modifications to focus on the younger eighteen- to thirty-four-year-old segment of its audience, the managers at WBCN now turned to reassess the soundness of their air talent. "WBCN had been as important to Boston as the *Globe* and the Red Sox; we were an institution," Oedipus observed. "But now, nineteen- and twenty-year-olds could no longer relate; they wanted something else." The program director referred ominously to the sacrosanct "Big Three," intact and in place on 'BCN's weekday air since 1980. At the moment, Laquidara and Parenteau were winning their respective battles, even with the younger listeners, but a shadow of scrutiny had now fallen on Shelton's shift. "Oedipus called me one day and said, 'I need to talk to you after your show,' the midday man recounted. "He said, 'You've been great and we love you. It's been such a perfect fit, gliding from Charles's insanity to Parenteau's wildness, the calm between the storms. But the company just bought WZLX, and there's a big master plan; there's going to be lots of changes.'" Shelton suspected what was coming next: that "master plan" probably included moving some or all of the 'BCN old-timers over to the classic rock station. "Oedipus said, 'You're the highest-paid midday guy in America. Your contract is up in a year; you're welcome to stay and keep doing what you're doing. [But] Mel loves you; he thinks you could be a [good] morning man. How about doing mornings at 'ZLX, with more money?'"

Shelton maintains that his relationship with Oedipus and Tony Berardini had curdled by that point over a union matter concerning insurance coverage: "They hated me and I hated them." Oedipus, however, saw it as strictly business: "It happened to Shelton: for an eighteen-, nineteen-, or twenty-year-old [listener], he was now their dad's DJ."

"They would have fired me right away if I didn't have the guardian angel, Mel Karmazin, sitting on my shoulder," Shelton countered. The jock walked away from his long-standing midday home in July of '93 to assume morning drive-time duties at Infinity's new classic rock acquisition. Despite the bad blood that had spread between the players in this episode, the move was strategically sound and mutually beneficial. The older WBCN listeners who loved "Captain Ken" for all those years would now follow him over to WZLX, where he played the legendary songs they enjoyed, while Oedipus was now free to replace him with someone he saw as a better and younger fit for the station's changing image (currently under construction). The ironic element was that Shelton now competed for listeners in the same time slot as his buddy Laquidara, the radio maniac he had clowned around with during "Mishegas" crossovers for well over a decade.

With Shelton's departure/ouster, Bradley Jay, a veteran of various shifts at WBCN since 1982, moved into the midday slot. Although now a twelve-year veteran of the station, the DJ possessed the hunger and image of a much younger and savvy new-music-oriented jock. With his unquestioned enthusiasm, Jay had proved to be resilient in whatever role Oedipus asked him to perform, from famously clowning around backstage with David Bowie and hosting Lunchtime Concerts in the eighties to playing a controversial figure in a salacious evening show before Howard Stern replaced him. Jay

Bradley Jay and Tami Heide get close for the camera.
Photo by Roger Gordy.

had moved into the night shift in '92 with the challenge of trying to head off the slide of listeners, not only from 'BCN, but also from radio in general, which occurred right after drive time. "Oedipus said to me, 'I want you to go for it, do things to make people notice,'" the jock remembered. In the quest for ratings, perhaps inspired by the success of Stern's show in other markets, WBCN went where it had never gone before: into R-rated territory. "We called the show 'The Sex Palace'; it was the Howard Stern show, but Howard wasn't there yet. We had strippers dancing in the window on Boylston Street, stopping traffic!" The lascivious spectacle, described on the air in every detail, drew a large crowd of listeners and pedestrians, who might have been stunned at the tawdriness of the moment yet remained to gawk at all the undulation, where, just eleven years earlier, Tony Berardini and Marc Miller had solemnly addressed hundreds of John Lennon mourners.

"'The Sex Palace,' was a little 'out there,'" Oedipus admitted. "It worked for a while, but Bradley just couldn't make the show broad enough, intriguing and risqué enough, to draw in a large audience like Howard Stern. [He] couldn't quite pull it in." When the program director and Tony Berardini took the monumental step of importing the actual master of that game, Jay was left without his full-time shift. "I didn't freak out, I didn't get bitter, I didn't burn bridges." He left for Los Angeles but only stayed away for a few weeks before returning to do more part-time shifts. But during that brief period away, the jock redefined his radio approach. "I read this book on marketing that said if you have a product, like mouthwash, and there's already a Listerine on the market, don't make another product that 'kills germs'; make it different. There was already a Howard Stern, so it was pointless to be a shock-jock; I did the exact opposite and went minimal." As the buffer between the up-tempo mischief of morning and afternoon drive times, the music orientation of the 10:00 a.m. to 2:00 p.m. shift had always been the correct recipe for success at 'BCN, and Jay poured himself willingly into the mold. "That seemed to be attractive [to Oedipus] because that is how I pretty much eased into the midday."

The next step was the music itself. Never a stranger to presenting the latest records and new bands, WBCN by default had always featured an ample variety of new music. By mid-1992, the station regularly played the music set free by Nirvana's hydrogen-bomb-like arrival: Seattle "grunge rock" and its major purveyors, Pearl Jam, Alice in Chains, and Soundgarden. The later, more commercial "postgrunge" bands such as Stone Temple Pilots and Bush found a home at WBCN, as did the "Brit pop" groups like Oasis and Blur. These exciting new sounds were featured freely next to mainstream acts like the Rolling Stones, The Who, and Pink Floyd. "We were playing the stuff we loved," Bill Abbate mentioned, "Pearl Jam instead of the latest from Huey Lewis and the News. It was new, it was fresh, and it was fun." In 1997, the industry tip sheet, *MQB*—*Modern Quarterback*, quoted Oedipus about the change: "'BCN had been defending its upper demos from 'ZLX by leaning hard on its classic rock library. 'It was time to let the classic rock station succeed on its own terms; there was no reason to fight them anymore.'" Steve Strick (who had become assistant music director by this time) agreed: "We already had a classic rock station in town; why go against them and split an audience that was not growing, but perhaps, dwindling in size? Everybody in the room was leaning toward the younger, more modern route. It was a more exciting direction to go, and even Tony went along with it, because he saw it as a way for the station to evolve and grow. We decided to do it gradually: playing the new music and starting to weed out the artists who were incompatible." Strick smiled and added, "So every week [in the music meeting], there was this fun little exercise of eliminating another artist. One of the first to go was Jethro Tull. Everybody was jumping up and down, 'Great! Snot dripping down his nose—get that

Steve Miller in the studio with Mark Parenteau, also producer Jeff Myerow. Changing times and a younger-oriented format soon leaves artists like Miller behind at WBCN. Photo by Mim Michelove.

off!' We stuck with a few, but it took about six months to rid the station of the older stuff."

A WBCN playlist from June 1993 included new music artists—Stone Temple Pilots, Porno for Pyros, Radiohead, Frank Black, Alice in Chains, and Soul Asylum—but was dominated by the latest singles from classic names like Rod Stewart, the Steve Miller Band, Pat Benatar, Donald Fagan, Pete Townshend, Van Morrison, and Neil Young. Nearly two years later, in April 1995, another sample playlist revealed that 'BCN's transition had been completed: artists such as Green Day, Live, Offspring, Morphine, Filter, Oasis, Pearl Jam, Belly, and Matthew Sweet completely commanded the roster, the sole classic entrant being Tom Petty and his song "It's Good to Be King." This development prompted Jim Sullivan in a May 1995 *Boston Globe* article entitled "Reinventing the Rock of Boston" to declare, "AOR is dead. Or dying. Or mutating. Oedipus wants nothing to do with those three now-dirty letters. 'WBCN,' he says, 'is a "modern rock" station.'" Sullivan also pointed out that this shift to play alternative sounds was a national trend, with several of the approximately 175 album-oriented rock (AOR) radio stations making the switch to play alternative music and join the 64 or so modern rock stations on the other side. "'We're redefining the center,' says Oedipus. 'The center will not hold; it's very askew.'"

In Sullivan's article, Oedipus admitted that the transition from full-on AOR to a modern rock entity took well over a year to complete, but the experience of attending Woodstock '94 "gave it a big kick." That festival, a twenty-fifth-anniversary tribute to the original three days of peace, love, and music that helped mightily to transform a generation, was held 12–14 August 1994 in Saugerties, New York. Santana, Joe Cocker, CSN, and John Sebastian returned to encore their now-mythical performances from the first concert, joining other classic rockers like Bob Dylan, the Allman Brothers Band, Traffic, and Aerosmith. The cream of the modern rock movement also mounted the two main stages at Woodstock: Nine Inch Nails, Green Day, Red Hot Chili Peppers, the Cranberries, and many more.

As a player during the original festival and still in business to enjoy the vibes a quarter century later, WBCN's presence in upstate New York was a must. "'BCN had the foresight to rent this house, a kind of cheesy ranch house which, for some reason, smelled horribly like fish," Bradley Jay remembered. "You could hear the bands in the distance, and you could walk there." The place ended up being called "The Love Shack," and the crew from WBCN used the house as a base for its broadcast operations during the rainy weekend. "I was there early, but the day the show started, people started to roll in. Oedipus had decided that we were a station of the people, and he made sure we broadcast that everyone should stop by 'The Love Shack' on their way to the festival. So they came all the time, even in the middle of the night, like zombies! They'd be knocking at the door, [saying] 'We heard on the radio we could stay here.'"

"We got inundated with people," Albert O agreed. "Parenteau suddenly showed up, John Garabedian was in there, J.J. Jackson, this woman Kat who worked at 'FNX at the time."

"It truly was 'The Love Shack,'" Jay commented. "At one point I got up in the middle of the night to use the bathroom, and it was like a Monty Python skit: every room, every bunk, everything, and everywhere: sex was going on." Vegas rules applied: what happened at "The Love Shack" stayed at "The Love Shack." Jay added, "It was good times there, till the people from MTV came in to use the bathroom and clogged it up!"

Never buoyed by an explosive political undercurrent, nor a joining of voices against an unjust war, nor a historic gathering of "freaks," the Woodstock '94 festival nevertheless contained some terrific musical performances and assembled a massive number (over three hundred thousand) of WBCN's revised target demographic: eighteen- to thirty-four-year-olds. For Oedipus, the experience proved to be as vital as it was enjoyable, serving as a research project in a gargantuan test market. Thousands stood in what became a muddy mess as the skies constantly hemorrhaged, but the overwhelming preference of the crowd for this new music confirmed to the program director that the fresh direction of the station was the correct one to take. Oedipus marched back from Woodstock all fired up. He was waiting for Steve Strick and me when we got into 'BCN to begin our workweek. Breathlessly, he described his experience and then began discussing whether we should accelerate the process of removing classic artists and adding modern ones to the playlist. We decided to experiment by going 100 percent modern rock on the weekends, a gutsy move considering that there were no focus groups, no strategic studies, and no call-out research to add to our gut instincts. But, within weeks, the ratings on those Saturdays and Sundays had improved so much, so fast, that we had to look at each and admit, "What were we waiting for?"

"We are definitely the New Woodstock station," Oedipus told Jim Sullivan in his '95 *Boston Globe* piece. "WZLX is the Old Woodstock station." The article went on to acknowledge that in a half year of programming modern rock, WBCN had lost ground to WZLX in the 25–54 audience, but (as intended) "in the younger 18–34 male category, [WBCN] significantly increased listenership over WZLX and WAAF."

With Ken Shelton gone, Bradley Jay in the midday seat, Howard Stern on at night, and modern rock the format, 'BCN headed into 1995 as a reinvigorated fighting machine. But the new music, as refreshing as it might be to the air staff, came to the station with a price. Because it was key to create hits out of these new songs, anchoring the format with the most powerful and coveted tracks, WBCN began playing its strongest music in a rotation of five to six "spins" a day. With that, the opportunities for the DJs to add their own optional songs (a tradition of varying freedom over the years) had ended completely. Now, the only musical freedom that could be exercised by an individual member of the air staff was if he or she attended the music meeting and voiced an opinion. "It was a slow evolving of complete freedom to gradual acceptance, and by '95 it was really tight," acknowledged Mark Parenteau. "I had started out way back making no money with 100 percent freedom; now I was making a quarter-million dollars [a year] and I had *no* freedom. But, it was a trade-off I gladly gave them." Most of the jocks, though not making anything near Parenteau's pay scale, had to agree that it was much more fun playing the music they enjoyed the most, even if the ability to choose individual tracks during their shifts had been curtailed. "Lots of people, from U2 to the new artists, were turning out great records," Steve Strick said. "It was a no-brainer to tap into that youthful energy." It also resonated with WBCN's past, which had always embraced the new and revolutionary, the up-and-coming, and the future trendsetters. "A roster once clogged with Classic Rock bands (Led Zeppelin, Pink Floyd) now churns out a steady diet of new-rock acts such as Belly, Liz Phair, Oasis and Bush," Dean Johnson wrote in the *Boston Herald* in May '95. He concluded, "There was a hole in the Boston market for a new-rock/alternative music station with a major FM signal. There is no hole anymore—and WBCN will reap the rewards."

ANY GIVEN SUNDAY, ANY GIVEN WEEKDAY

The WBCN press release dated 17 November 1994 announced the startling news, but the local media was already buzzing excitedly about it: WBCN and the New England Patriots had signed a three-year deal for the radio broadcast rights beginning in the 1995–1996 season, ending the football team's previous relationship with WBZ-AM. The release stated, "Infinity's latest addition will mark the first time a major league sports team will be broadcast on an FM station in stereo in the Boston market." The news broke on "The Big Mattress" as Robert Kraft, who had recently become owner, president, and chief executive of the Patriots, and WBCN general manager/vice president Tony Berardini shared the announcement over the air.

The Kraft family, which had made a fortune in the paper and packaging materials business, had purchased the team early in 1994, and it was an association with Robert's son, Jonathan, that eventually led to his family's involvement with WBCN. Tank remembered, "On Fridays I would broadcast my sports reports from outside the station on the sidewalk. We had clients involved, restaurants coming down, caterers, bicycle tune-ups; I knew Jonathan because he'd come down too. Once day, I introduced him to Berardini and suggested (ha, ha) that they talk about getting the Patriots on 'BCN, not that that was ever going to happen, of course!"

"Jonathan used to come in and sit in on Charles's show all the time," Berardini recalled. "He loved 'BCN. One day Tank introduced us and we talked. We went out to dinner, and Jonathan made it very clear to me that they wanted the Patriots to be a success. They had been season ticketholders since Jonathan was a little kid, all those years when the Patriots couldn't even get fifteen thousand fans in the stands—sitting on those metal benches. And you have to remember that after '86, the Patriots just sucked; they were bad after they got blown out by Chicago [in the Super Bowl]."

With the ink barely dry on the new owner's contract to purchase the team, Berardini initiated a meeting with Mel Karmazin, Jonathan, and his father Robert Kraft. Berardini explained,

We had it in Robert's office, and I was shaking like a leaf, totally excited, but nervous. It was supposed to be an hour meeting, and Jonathan and I actually said very little the whole time. Mel and Robert started talking, and it got into, "Hey, do you know this guy?" They were talking about different investment bankers, because by this time the Krafts had a huge business, so Mel and Robert knew a lot of the same people. I was looking at my watch; we were forty-five minutes into this, and they were still playing "What banker do you know?" But then Robert started asking Mel about the radio business. It wasn't negotiating or anything; he just wanted to know about the business. Mel started telling him about how you make money in radio, the performance of Infinity over the years, that we were a growth company, the expectations, this and that. All of a sudden, Robert told us, "I want you to meet somebody." He picked up the phone and said, "Hey, can you come up here?" He introduced us to this guy who was in charge of investing funds for the Kraft group, and then told him, "This is Mel Karmazin, the CEO of Infinity. I want you to buy stock in his company." Mel looked at Robert and said, "I came here to do a deal for the radio rights to the Patriots, not to sell stock!" But it struck me that this was Robert's way of saying, "I like your business, and I like the way you *do* business."

Mel Karmazin made his pitch to the Krafts after that, and as usual, according to Berardini, he kept things simple and direct:

"You should be on FM radio and here's why. We can do things in stereo; it's a much cleaner signal and we can mount as good a production as an AM station. If you're worried about the signal coverage, we will go out, just like we did in Dallas, and get affiliates in every town in New England who would love to have the Patriots broadcast." Then Mel said, "Let's talk about the money." Robert looked at him and said, "Why don't *you* tell me what *you* want to do?" Mel came back with something like, "I don't believe in going back and forth; I don't want to get into negotiating against WBZ [which was looking to renew its contract], so you tell me what your number is and then we'll tell you whether we can do it or not." Robert still wanted it the other way, but Mel said, "No. If you give me the number, I'm going to turn to Tony and say to him, 'Tony, you can't lose money on this deal; can you make this revenue?' And Tony is either going to say 'Yes' or 'No.' If he says 'No,' then tell 'BZ I offered twice as much as I did, and then you get it from them."

I was sitting there going, "Oh shit. What's this number going to be?" I was trying to calculate everything in my head because I didn't want to say yes if I couldn't do it; there would be hell to pay. They went back and forth, but finally, we concluded, "We'd really love to do it, but why don't you guys think about it." Mel and I walked out, and he said, "I don't know; we'll see what happens." He got in his car, went to the airport and back to New York. I drove to the station. I knew what money they were paying in Dallas and Philly . . . big money. But I really wanted to do it. So I called Jonathan and said, "Okay, those guys did their thing. Let's you and I figure out what the number is going to be." So, we batted back and forth. Then I called up Mel and told him I had talked to Jonathan and had the number.

"Can you do it?"

"Yes."

Tank related what happened next: "So, then, one day I'm on the air with Charles, and Jonathan walked into the studio with his dad. When they broke the news, I went crazy! I was like Meg Ryan in *When Harry Met Sally* when I heard we got the rights to broadcast the Patriots. That was so cool; I just went nuts: 'Yes! Yes! Yes!'" Moments later, astonished and exhilarated, Tank called Bill Abbate, who had just put in a long overnight shift leading into "The Big Mattress" and since gone home, missing the surprise announcement. "I started to fall asleep and the phone rings," he told *Virtually Alternative* in 1998. "It was Tank, who's screaming, all excited, 'I can't believe it; we got the Patriots.' Tank and I were ecstatic, 'cause we were both sports fans; we really had no idea where this was gonna take us, and as it turned out, he and I were cohosts for the pre- and postgame show for the '95 season, our inaugural season. The team only went six and ten that year; it was kind of rough. But I'll tell you, being one of only thirty sets of broadcasters working in the United States in the NFL was a lot of fun."

"The Patriots were a laughing stock at first, and we stepped up big time, paying considerably more than what had been paid in radio rights beforehand," Berardini summarized. "That was based on a belief that the team was going to be successful because of the type of ownership that existed in Foxboro. I just believed in the Krafts."

Curtis Raymond, employed at WBCN as retail sales manager, was initially tasked with finding someone to be in charge of selling time for the games but ultimately took the responsibility on himself. The new sales manager for the Patriots broadcasts remembered, "This was the first [time that football was] on a rock and roll radio station; it caught the Boston media totally off guard. I can remember all the agency experts saying it would never work; 'blah, blah, blah.' But it did." Critics thought that WBCN's trademark music programming would clash with the mostly talk world of sports broadcasting, but Berardini disagreed that a conflict existed. "There were some worries about whether or not it would be right for the radio station, but football was compatible with the male demographic."

"'Nocturnal Emissions' was on Sunday night; it's not a popular time for radio listening," Oedipus observed. "So, Sundays, we figured, 'let's give it up; it will generate a lot of revenue and bring our radio station a lot of audience, who will leave their radios tuned to WBCN [after the game]. Then, on Monday mornings, the listeners will be back when they get up and go to work.' I don't regret that decision at all."

"We took a few hours to broadcast something that everyone, at first, wasn't a part of," reasoned Bill Abbate, who would now double as a DJ and sports announcer. "But by the time of my last broadcast in February 2004, it was our third Super Bowl, and the people were certainly there!"

WBCN gets the Patriots broadcast rights and Bill Abbate happily does double duty as DJ and weekend sports announcer. Courtesy of WBCN.

"It's an expensive proposition, but sports, more often than not, are a real benefit to a station, even if you can only break even," Curtis Raymond explained. "I use the example of television: when FOX got the NFL, nobody had heard of *The Simpsons* and *Married with Children*, but they were all promoted on the football broadcasts. The same thing held true for radio. We didn't break even the first year, but we had some pretty spectacular weeks in the years after. One of the best was the year after they went to the Super Bowl with Drew Bledsoe as quarterback [in the '96 season, losing to Green Bay]; we were second or third in the country in football revenue that year." Raymond got to work establishing the "Patriots Rock Radio Network," extending the reach of the games past the limits of 'BCN's radio signal, which as an FM transmission, paled next to the range of the AM band. "WBZ, being the predecessor, had that clear channel signal and didn't need as many affiliates; we had to get a bunch of new ones," he explained. "In the first year we were in the forties [in stations], as far down as Danbury, Connecticut. I remember Stephen King signing the affiliate agreement for [his station] 'The Zone' [WZON] in Bangor, Maine. I hung onto that agreement!"

During the actual games, Gil Santos and Gino Cappelletti, the already well-established team who had worked the booth for WBZ, called the play-by-play, while Oedipus and Berardini eschewed hiring other sportscasters, to work in-house and let Tank and Abbate handle the pre- and postgame segments. The August 1995 debut of WBCN as the flagship station for the Patriots' radio broadcast marked the end of nearly a year of preparation, yet the first game against the Detroit Lions was not without its hiccups. Veteran *Boston Globe* sportswriter Jack Craig attacked the station mercilessly in his article entitled "WBCN Drops Ball in Debut," calling the broadcast "woeful at times" and the two-hour pregame show with Tank and Abbate "like amateur night." Blatantly unimpressed, the writer also commented that "the pre- and postgame hyperbole and bubbly manner sounded more like a lead-in to a rock concert than a football game." He even castigated the station for its weaker signal. About the only thing Craig did approve of was the focused chemistry of Santos and Cappelletti, who "kept things in perspective during the game. No foolish praise fell from their lips." As the play-by-play tag team remained the only element of the broadcast held over from previous years, the writer, not so subtly, gave the radio station a parting "face-mask" in his assessment. WBCN broadcast producer

Marc Cappello, then a nineteen-year-old promotions department intern hired to assist Abbate by locating Patriots stats during the games as well as recording press conferences, recalled, "I think a lot of people looked at us and thought, 'What do these idiots know about putting on a sports broadcast?' WBCN was a rock station, not a sports station; it was all so new to everybody. But Oedipus didn't panic; he said, 'Look, we're going to do our own thing, and we'll get better at it,' and we did."

If the stodgy Jack Craig represented the view of a typical football traditionalist, then he was going to be in for more surprises. Curtis Raymond remembered,

> Oedipus and Tony wanted cutting-edge promotions with the football games, so we did one once with a condom company. It was in the two-minute warning at the end of the game: if the Patriots *scored* in those final two minutes, then whoever *entered*, won a year's supply of condoms, plus tickets to the next game. When we presented that to Gil and Gino, there was some real trepidation; they weren't really comfortable with that. But then, about a month later and unbeknownst to us, that [idea] won the Radio Promotion of the Year [award] at *Brand Week* [magazine], got written up in the *Boston Herald*, and suddenly everyone loved it. With us, the promotion kind of made sense, so Gil and Gino did it.

Touchdown!

The production and professionalism of WBCN's radio broadcasts would advance dramatically, keeping pace with the Patriots' own improvements to its organization and the inception of a winning spirit. "The Krafts really had a commitment to making the team a success and making them part of the community," Berardini explained. "That's why we eventually did a second deal and a third deal, and the third one was for ten years! Who knew that was going to happen?" Although Tank would exit WBCN with Charles Laquidara, regretfully leaving the Patriots Rock Radio Network after one season, Abbate carried on with new cohost Mike Ruth for the '96 season, one in which the Patriots went all the way to the Super Bowl behind the coaching of Bill Parcells. "Here I am," Abbate told *Virtually Alternative*, "two seasons working the National Football League, now I'm broadcasting live from the Louisiana Superdome for Super Bowl XXXI!" Although the team lost that contest, future steps for the Patriots would cement their legendary

Hard work pays off: Tony Berardini with Jonathan Kraft and Robert Kraft
holding the Super Bowl trophy at WBCN in February 2002. Courtesy
of Tony Berardini/New England Patriots.

status. The unit, under the direction of new coach Bill Belichick, advanced
to the Super Bowl at the end of the 2001 season and defeated the St. Louis
Rams to become NFL champions for the first time.

The Super Bowl in New Orleans became a great moment for the station, not only because of the Patriots victory, but also because the game's
halftime show featured the group that WBCN had stood behind since its
fledgling night playing the Paradise Theater in 1980. Performing on entertainment's most-watched and prestigious stage possible, U2, at the height of
its own game, turned in a heartfelt and emotional tribute to the victims of
the 9/11 attacks. As part of his assignment, Bill Abbate attended the earlier
press conference with the band, hopefully to tape some excerpts to present
on his pregame show. "It was in this gigantic room with all these people
and I managed to ask U2 a question," he recalled. Once Abbate's flailing
arm had been recognized, he had the floor:

> I gave it this big buildup, mentioning 'BCN, kind of like, "We're from Boston,
> the city that helped you get your big start here in North America, the city
> that's got the largest Irish population outside of Ireland."

"I know where this is going," Bono says.

"Do you have a rooting interest in the game?"

"I'm not the guy who knows anything about football; let me pass this one over to the Edge." So Edge goes, "Well, we really don't know much about football"; then for the next sixty seconds goes on and on with specific differences between the Patriots' defenses and the St. Louis offense and defense, how Brady is more of a pocket quarterback while Warner can run around and scramble. The whole crowd was just rolling! "But we don't know anything about football." Yeah, right!

WBCN looked and felt a lot different from its Camelot years, with football now a part of the broadcast week, the taped Howard Stern show on every weeknight, Bradley Jay assuming Ken Shelton's midday dynasty, and the radical transformation of the playlist. Now classified as a "modern rock" station by *Billboard* and the other music trade publications, 'BCN focused on current alternative music choices like Foo Fighters, Weezer, and Alanis Morissette. Because of, or despite, these significant adjustments, the station continued to maintain the upper hand in ratings and revenue success, focusing on the younger male demographic and mostly abandoning the older twenty-five to fifty-four crowd to 'ZLX. By the spring of '96, WBCN stood at a ranking of number 6 (tied with WZLX) for all Boston listeners twelve years and older. By contrast, WAAF took number 14 with a 2.6, and low-power 'FNX came in at number 19 with a 1.6. Having departed the station by this point, marketing and promotions guru David Bieber later observed, "WAAF was around at that point [the station had moved its studios from Worcester to Westborough], and 'FNX had its own valid mission, but it was an effort just to get 101.7 [in on the radio]. I think both of those stations deserve respect, but 'BCN was just a behemoth."

As the reigning patriarch on WBCN, Charles Laquidara could have easily been viewed as an anachronism amid the station's new fascination with a targeted eighteen- to thirty-four-year-old demographic. But his "Big Mattress," now in place for over twenty-two years, continued to garner respectable ratings, and his attitude remained as fresh as ever. "I was very aware that times were changing," he said, "just like I knew the times were changing back in the sixties. I didn't want to be like my parents or my grandparents and be one of those people who didn't let go. When the change comes, you got to deal with it." Laquidara's long-running music spotlight, "The Big Mattress Song of the Week," reflected his embrace of the new music, as did the DJ's efforts to learn about the latest groups he featured. "Oedipus kind of made fun of me because, on one of my trips to Hawaii, I was trying to memorize the names of all these new bands, like Collective Soul and Our Lady Peace, and the names of lead singers and guitar players. I actually had lists of these bands that I was memorizing on the plane. When I came back and told him I had done this, he just laughed at me and said, 'You don't have to know the names of these guys; it's not like Charlie Watts from the Stones or the Edge from U2. All this music and all these groups are disposable.' Things were changing so fast."

Despite the cultural box spring shifting beneath it, "The Big Mattress" itself remained firm and inviting to listeners. Laquidara, as was his norm, continued to surround himself with a talented staff: writer, occasional co-host, and resident cynic (the show's "Bitter Man") Lance Norris; surfer-dude newsman Patrick Murray; producer Bob Malatesta; writer and voice of reason Don Bertolino; longtime veteran Tank ("The Round Man"); and a herd of others. But "The Big Mattress" did have its detractors, and they emerged largely because of the *other* morning show that was on WBCN at the time. "I was constantly faced with these people who wanted Stern on in the morning and wanted me out of there. These people really did have an ageist attitude: 'You're old; you're in the way.' Howard, as nice a guy as he was in real life, and as nice as he always was to me on the radio, never treated me with any disrespect—like he did with other people. I mean, he'd have funerals for jocks when he moved into a town! He never treated me like that. But his listeners were not so kind; they'd go after me, just like in the old days when 'COZ listeners would leave their beer cans in my driveway."

When Infinity first made Howard Stern available to WBCN, Laquidara had been defended by his managers. "At that point Charles was just killing it," mentioned Tony Berardini. "We looked at Mel and said it wasn't going to work with Charles doing so well. So we got Howard to do nights." Karmazin didn't push it; he knew how beloved and legendary Laquidara was in Boston, and he had the ratings to prove it. Years later, in a *Globe* retrospective from 2000, Jim Sullivan would write, "Mel Karmazin lovingly calls him a 'huge pain in the butt. I met him in 1981 and I've been a fan ever since. He's a character, and I think radio needs more of these.'" In fact, the DJ had always enjoyed a friendship with Karmazin from the earliest days of Infinity. Tickets could have been sold for people to sit courtside and watch

Laquidara, the classic liberal, line up against the conservative corporate CEO, to publicly spar on every subject from raising children to apartheid. These good-natured debates, with edgy punch lines zinging left and right, were often fueled by deadly serious concerns, but mutual respect always held the two in close rapport. Nevertheless, as Tony Berardini described, "When [Stern's] next contract came up, Mel came to Oedipus and I and said, 'Part of the way I got Howard to resign with Infinity was to promise him mornings in Philadelphia and Boston. So it's up to you guys; you can have him in the mornings if you want to. If not, Howard is free to negotiate with another station in the market.' Mel outlined the situation and left it up to us."

But, it was quite obvious that Tony Berardini and Oedipus had been thrust into a catch-22 situation. To lose Stern would give the ratings star a foothold elsewhere to deflate "The Big Mattress" and 'BCN as a result. "In every market Howard had gone into, he destroyed the competition," Berardini observed, "and Charles would have been his number 1 competition." The alternative was to keep Stern and move him, live from New York City, into the weekday morning slot. But where would that leave Laquidara? This scenario meant that the market veteran would exit his job and, perhaps, the company. Nobody liked this option; WBCN's star personality remained on a first-name basis with most of Boston, and to lose that relationship and market recognition because of collateral damage just didn't make sense.

"I know Oedipus had the pressure of not having Howard Stern on in the morning and not getting those great ratings; I think everyone knew that," Laquidara admitted. "I was getting good ratings myself, but I had my ups and downs." Oedipus conceded, "Howard fit with the audience we were going after, plus he dominated the ratings, and it would have *dramatically* hurt our radio station if he wasn't there. This would have affected salaries, jobs . . . everybody [at WBCN]. If Howard had gone to 'ZLX, for example, we would have gotten killed."

"We had a meeting," Berardini described, "Mel, Oedipus, and myself, in my office, for a very long time. We were racking our brains trying to figure out what we could do."

"It wasn't easy," Oedipus admitted. "I saw Charles every morning, more than I saw my own family. Charles was . . . my big brother." With every go-around in the discussion, the two fates remained resolute: WBCN would either be crippled or wildly successful, and Laquidara would be hurt either way.

"I don't know who made the suggestion," Berardini said, "but among the three of us, we hatched the idea: what if we put Howard on in the morning at 'BCN and moved Charles over to WZLX because they needed a better morning show?" The thought hung between them for a moment. "I remember Oedipus sitting there; he got a look of relief on his face, and he said, 'You know, that could work for both WBCN and WZLX, and it could work for Charles.' But we were really concerned; we didn't want to move him. How do you talk to a guy who's been a legend in the market since 1969? 'By the way, you have to leave the station that you helped start, to go across the street.' Oedipus said, 'Let me talk to Charles; I can convince him this is the right move.'"

"I didn't see it coming; I wish that I had," recalled Laquidara. "Oedipus asked me to meet him at the Providence Restaurant [in Brookline]; he wanted to chitchat a bit. So, I thought it was a regular meeting. But when I got there he had three glasses of my drink of favor at the time, single-malt scotch, lined up on the bar waiting for me. Uh-oh. 'Have a drink,' he said, and then he laid it on me."

"We had an intense talk about where radio, WBCN, and his career were going," Oedipus continued. "I said, 'If you don't move and Howard goes on a different radio station, you'll go out in a blaze of glory, but you'll fade quickly.'"

"I don't think Oedipus realized he was doing it, but he set me up," Laquidara added.

He said, "Now, you don't have to do this; you don't have to go to 'ZLX if you don't want to. But if you don't go, Howard will kill you on the radio, and I won't be able to protect you." So, what he was saying was, "Charles, you can stay here at WBCN and you have a strong following of listeners who would support you if you took that stand, but you would be going against the company. Howard should be on in the morning and most of the people who like classic rock are over on WZLX; you should take the high road and do what's right for the company and what's right for the listeners, and just go." Logically, he was right, but emotionally, it was devastating to hear. So, I tried to be cool about it and take that high road.

"Charles ultimately made the right decision; he went to 'ZLX, took his audience with him, and he was very successful," Oedipus summarized. "And that was not only Charles's audience; it was his music and his era, no matter how much he claims to like new music. He made the right move by no longer trying to prolong his career [at 'BCN], and was able to leave on his own terms." Laquidara came up with the idea of doing [the switch] on 1 April: "Because, you know, every time I got fired, whether it was by Charlie Kendall, Klee Dobra, Arnie Ginsburg, or whoever, it always seemed to be on April Fool's Day!"

Bright and early on 1 April 1996, fans of "The Big Mattress" were surprised to hear Howard Stern on WBCN broadcasting loud and clear, live from New York, while their hero sat in for a displaced George Taylor Morris (who had bumped Ken Shelton the year before) over at WZLX. An inspired joke, many would conclude, but most didn't suspect that, when the next morning dawned, Laquidara would still be absent from his longtime post and 'BCN would actually have its new and permanent morning man in place. Tony Berardini said, "The way we were able to orchestrate it made it good radio, good theater." The stunt succeeded in generating one of the biggest buzzes in Boston radio history, absolutely focusing the city's attention on the momentous swap. Susan Bickelhaupt at the *Boston Globe* broke the news early on 2 April: "The change is expected to be a permanent one, to be announced at a press conference this morning via satellite from Stern's New York studios."

"There were listeners that were outraged," Oedipus told *Virtually Alternative* in 1998, "but it was like, if you want Charles Laquidara, you know where to get him; and if you want Howard Stern, you know where to get him." In his book *The More You Watch, the Less You Know*, the always-opinionated Danny Schechter gave his own estimation of the switch: "The station's longtime soul and defining spirit, my former partner in crime, the 'morning mishegas' man Charles Laquidara, who invented many of the on-air shticks that Stern, as a BU student and early BCN listener, appropriated, was ordered into exile at another Infinity-owned Boston station. Charles was hurt, but well-compensated."

At WBCN, meanwhile, Stern's move to a timeslot he shared on "25 other stations," the *Globe* reported, meant a cash windfall as advertising rates soared and the sheer number of commercials on the air in morning drive increased dramatically. Oedipus said in *Virtually Alternative*, "It was clear that not only did he belong in the mornings, because that's what he does, but also he could generate a lot more money in the morning." So strong was Stern's appeal, listeners would sit through a hitherto forbidden number

Bumped by Howard Stern, "The Big Mattress" finds a new bed at WZLX. Charles Laquidara in happier times (circa 1988). Photo by Mim Michelove.

of commercial spots while they waited patiently for the programming content to return. How long? The *Washington Post* reported that "ad breaks on the Stern show have been clocked at as long as 18 minutes, 48 seconds, with as many as 30 separate spots jammed together." Many advertisers would have nothing to do with the raunchy show, but sales manager Bob Mendelsohn said there were plenty of others more than willing to step up: "It opened a door to a whole other universe of people who kind of came from the same school. Advertisers realized, 'Hey, those people who listen to Stern because of all the smutty talk? Those are my customers.'" Young, white males, in particular, were attracted in droves to the show. "We sold a lot of Met-Rx!" Mendelsohn laughed, referring to the popular bodybuilding supplement gobbled up by grunting, young weight lifters.

"For a lot of those advertisers, they would pay any price that you asked, because we had separated ourselves from the pack; we had Howard Stern," Mendelsohn pointed out. "There were times that if we had sold out and someone really wanted to get on the air badly, we'd get two or three thousand dollars a commercial with Howard. They would pay that."

"It was a challenge selling Howard," Tony Berardini observed, "but we were able to find those sponsors that did buy [him]. It was almost like you had two completely different advertiser lists."

"It was very much a split deal," Mendelsohn agreed. "There were advertisers we had cultivated for years who said, 'No way!' to being on the air with Howard. It would drive us crazy mechanically: what do you do when, instead of ending at 10:00 or 10:15, Howard would decide to stay on till 11:30? That was a real problem. But there were others who didn't care; they just wanted the media value [of Stern] and that was the biggest thing to them."

Stern's presence was a vastly polarizing influence at WBCN: those entertained by the curly-haired and bespectacled radio host became staunch supporters, while detractors simply abhorred his approach. Oedipus had been a fan even before Karmazin offered Stern to 'BCN, and in his mind, the move to mornings did not alter his primary desire for the radio station: "My job was to keep [WBCN] vital, interesting, and exciting; plus, I still wanted to play all this new music. I wasn't worried about a talk show in the morning because Charles had already done a talk show there. We hadn't been breaking any new music in the morning; it was all about entertainment. And Howard *was* entertaining. It prolonged the radio station; 'BCN would have never lasted for as long as it did."

"I'm a big Howard Stern fan," Chachi Loprete confessed. "I enjoyed the content, not the dirty talk; I thought that no one could do an interview like Howard Stern. He could talk to anyone and make it sound interesting. But it was a double-edged sword; I hated to see Charles go. He was the reason that I started in radio." Others, like Danny Schechter, were worried or convinced that Stern's presence would forever destroy, or at least significantly alter, the fabric of what WBCN had always been in Boston. "The voice of the angry white man," he wrote in his book. "His tirades are rife with sexism, race-baiting, and homophobia, which help reinforce prejudices and encourage races and communities to oppose each other. His fans think it's funny."

To be honest, once Howard Stern had displaced Charles in the mornings, I was chagrined at my new, unwilling association with the infamous shock-jock and his supposed outrages over being challenged on the First Amendment right to free speech. I thought that was a thin veil to excuse just about anything that was licentious in his show. Although WBCN's positive mojo would hold me in place for another two years, from that point on I sought an opportunity to leave, after nearly nineteen years at 'BCN, and would do so in April '98 to join Charles at WZLX.

In the years that Stern would be on WBCN, from 1993 till he left Infinity for satellite radio in January 2006, the radio star never traveled to Boston to do a live show or participate in any local promotion, but the station would receive frequent visits from members of Stern's celebrity staff. Loprete mentioned, "I got to know Gary Dell'Abate [Stern's producer] very well; I was always hiring him to come in and do an appearance at a strip club or a store. I hired Jackie the Jokeman to do appearances." Then there was Crackhead Bob. "They called him that because he did too much crack and it made him mentally incapable; he could not communicate at all. We had to pick him up, take him to his hotel, check him in, make sure he had room service and a menu. Then he'd point out what he wanted to eat and I'd have to order for him. Then we'd have to pick him up and take him to his gig." So what did Crackhead Bob actually do for his pay check? "People would just want to meet him. I hate to say it, but he was sort of a sideshow act; he couldn't say words properly, so you had to really listen. I remember he did a gig in Quincy at a bar for two hours. There was no act; he just sat around and met people."

When Stern's movie *Private Parts* was released [in 1997], WBCN held a screening for it at the Nickelodeon Theater. A grand opening of a Block-buster Video store was tied in to the DVD release of the movie, with Dell 'Abate flying in along with Hank, the Angry Dwarf, another participant in Stern's cabinet of oddballs. "The Angry Dwarf was a major alcoholic," Loprete recalled, "and there was a huge line of people waiting to meet him. Then, this woman brought him a long dildo. He started, uh, working on it while people stood in line for autographs of the DVD. He didn't care. The manager of the store was disturbed about it. Who did she ask to get him to stop? Lucky me."

That's the pool Stern liked to swim in, but despite the blue tone, this was big business. Farid Suleman, Infinity's chief financial officer, told the *New York Times* in September 1995 that the shock-jock accounted for $14 million to $22 million of the company's revenue. Although he also stated that those figures represented, at the most, only eight cents of every dollar of Infinity's take, that deprecating comment might have had something to do with Karmazin's negotiating strategy, since Stern's contract at the time was just about up. But, for one personality to be responsible for that much percentage in any corporation was significant. At least Infinity thought enough of their star to shield him repeatedly against the FCC, which had begun to harass the jock relentlessly on obscenity issues; and that defense was certainly not cheap. Even one questionable on-air comment slipping out of the host studio and multiplied across two dozen affiliate frequencies, each with an accompanying fine, added up to astronomical totals. The *Times* article pointed out, "The company recently paid the Government $1.7 million to settle indecency charges against Mr. Stern for his comments on a range of topics like incest, masturbation and the unorthodox use of gerbils."

Defending Stern was all part of Mel Karmazin's big picture, and since any future deal would require approval of the FCC, Infinity needed to have its ducks in a row and be as unhindered as possible during any proposed expansion. He knew something that many citizens did not: in 1995 ongoing debate raged in Washington over whether or not to further deregulate the radio, television, and communications industry. Karmazin, a fearless gambler, had bet that the government would end up relaxing ownership rules again, allowing companies to purchase as many radio properties in a city as they wished. The radio boss figured right: President Clinton signed the Telecommunications Act of 1996 into law in February, and Infinity barely caught its breath before unleashing the dogs of commerce from its New York office. Geraldine Fabrikant stated in the 21 June 1996 issue of the *New York Times*, "In the last three months, Mr. Karmazin has been on an acquisition binge. Just last March, he agreed to buy 12 radio stations from Granum Holdings L.P. for $140 million." But, this was small potatoes; the article's main point concerned the stunning announcement that Infinity's president had masterminded a huge merger with Westinghouse Electric Corporation, which owned the CBS radio and television networks. Still subject to approval by the feds, the action, Fabrikant stated, "would create a radio giant with 83 stations and a constellation of radio stars from Howard Stern and Don Imus to Charles Osgood."

By the end of the year, the multi-billion-dollar agreement got its legal stamp of approval. Infinity merged into the CBS Radio Group, and Karmazin was named president of the whole, immensely expanded, shebang. It was not unlike being "kinged" in a game of checkers and then jumping over all of your opponent's pieces in one master stroke, except that the game board stretched nationwide across dozens of markets. Within just a few years, the Telecommunications Act of 1996 had opened the doors to wild expansion in the industry as most "mom and pop" radio stations were gobbled up and smaller radio chains became food for the hungrier corporate sharks. By 2003, the Associated Press noted, "The bill allowed Clear Channel Communications Inc., now the #1 owner of U.S. radio stations in the country, to grow from 43 radio stations in 1995 to more than 1,200 this year." The article also revealed that Infinity Broadcasting had enlarged to include more than 180 radio stations and that many of them "are located in the 50 largest radio markets, making them especially lucrative and high-profile."

Back in Boston, the station that had financed the initial stages of this empire soldiered on in its new alternative direction, with ratings powerhouse Howard Stern ensconced in AM drive. However, typical of stations that carried the New York shock-jock's show, WBCN now took on a somewhat bipolar personality. Tony Berardini explained, "Fundamentally, it changed the nature of WBCN. Howard was so unique that primarily it was his audience that listened to him, and then there were the others who listened to whatever else you were doing the rest of the day." Despite repeated attempts to recycle listeners out of the morning show to sample WBCN in its other day parts, the station soon became a schizophrenic entity with two nearly distinct and opposite audiences. Bruce Mittman, the general manager at WAAF, who had tangled with 'BCN for ascendancy on the radio hill for six years, quickly grasped this: "It was now Howard *and* WBCN. We saw this as an enormous opportunity, [because] the importance in building ratings is to be able to maintain an audience and recycle them into other day parts. 'BCN wasn't recycling. If we could further separate 'BCN into 'two stations' and attack the rock and roll half, we could win it all. The only way we could attack was to be as local as we could be. We had Greg Hill [WAAF mornings] everywhere that WBCN's morning show should have gone, thereby reinforcing the fact that Howard, although funny, was not local and that you couldn't relate to him; you couldn't touch him; [and] you couldn't interact with him."

The untouchable Stern brought WBCN enormous ratings and revenue success on a daily basis, but his presence had seriously unbalanced the radio station. The schizophrenia in its programming would not improve, at least not until the New York–based shock-jock solved the problem himself by opting out of terrestrial radio years down the road. Tony Berardini would later assess, "If you had to mark the day where WBCN fundamentally changed, beginning it on the path that eventually led to it going out of the format, our decision to put Howard on in the mornings was the turning point."

A BAD-BOY BUSINESS

WBCN's embrace of modern rock scored statistically from the first, but a dramatic visual confirmation of all this ratings hoopla played out in June 1995 when the station presented its first "River Rave," at Boston's Esplanade on the bank of the Charles. Four fledgling bands from the 'BCN playlist—Sleeper, Letters to Cleo, General Public, and Better Than Ezra—made up a modest bill barely expected to draw much more than five thousand people. That said, police presence was formidable, and the city was nervous about a repeat of the bloody riot that had occurred the previous year when WFNX presented Green Day at the same location. A beautiful spring day brought the sight of thousands of concertgoers streaming out of Back Bay and across the Storrow Drive pedestrian bridges, stepping off the Red Line at Charles Street and strolling over the river on Massachusetts Avenue. Like the brooms and their water buckets in *Fantasia*, the people just kept coming, until a crowd of over fifty thousand filtered onto the grounds for the free concert, dwarfing the stage and the small coterie of station staffers standing in amazement behind the security barriers. Worries of another melee could not be suppressed, and the police called in for reinforcements. "We were absolutely overwhelmed with how many people showed up," Roger Moore, WBCN's broadcast engineer for that day, related. "But, there were no incidents whatsoever; we were all expecting the worst and nothing bad happened at all."

The success of this first River Rave would inspire a new tradition to go along with WBCN's fresh direction, with the concert growing into an annual event. Oedipus commented, "The first one, at the Hatch Shell, was good, but the great Raves occurred later, the ones at Great Woods and then Foxboro." In future years, despite moving to these new locations that were nowhere near any river (if anything, only dirty culverts or muddy drainage ditches), the name stuck, and the River Rave became an annual expectation. Injected with steroids, the 1996 festival expanded to seventeen bands and relocated to Great Woods in Mansfield; the following year it morphed into a two-day extravaganza of thirty groups; and by 2000 the enormous digs of Foxboro Stadium were required to house the now-gargantuan affair. The unexpected success of the first event also emboldened Oedipus to create a Boston adaptation of LA sister station KROQ's "Almost Acoustic Christmas Show," calling it the WBCN "Christmas Rave." Two dozen groups and artists, showcased in eight venues, from tiny T.T. the Bears in Cambridge to the Orpheum Theater downtown, performed on one night in December 1995. There had been a precedent set when 'BCN staged Peter Wolf in an unplugged holiday party in the Middle East the year before, but the Christmas Rave was a full-blown electric event featuring future household names like the Dave Matthews Band, Goo Goo Dolls, 311, Jewel, and Ben Folds Five.

The Raves of 1995 underscored WBCN's format realignment, as did the presence of a new school of young DJs, like Shred, who had been at the station since 1988 but blossomed with the station's concentration on new groups and alternative sounds: "'BCN was always about breaking new bands; we made the hits. And it didn't have to be rock; even if it was a pop song and it sounded good, we'd play it anyway." Shred's favorite moments were meeting Damon Albarn of Blur; talking to Green Day in the 'BCN conference room; and speaking in awe backstage at Avalon with a personal hero, Johnny Cash. In '93 Oedipus hired petite redhead Melissa Teper for the weekends and fill-ins from a small station in Marshfield where she performed as the coquettish "Siobhan," playing tunes from the Emerald Isle. "DJ Melissa," as she became known, answered an ad looking for Listener Line volunteers on Charles Laquidara's show and, like so many before her, grabbed onto that dangling radio lifeline. "'BCN was great about that: giving the little guy a shot. That's what the station was all about: finding and nurturing new talent. Even Oedipus was an intern once." The wise-cracking Harrison came to the station in '94, joining the weekend air staff, and soon following Bill Abbate's example of doubling as a DJ and a member of the Patriots broadcast crew. Janet Egan, guitarist in the high-profile local metal band Malachite, worked at WFNX as a DJ and local music maven before Oedipus hired her away for similar duties at 'BCN. Under her radio pseudonym "Juanita the Scene Queen," she brought weighty street cred and experience to the lineup. Neal Robert defected from WFNX, after helming that station's afternoon shift for seven years, making the switch even though there were no full-time jobs to move into at his new home. "I still felt I had made the right decision; I was psyched to be at 'BCN."

By 1997, WAAF had been feuding on and off with 'BCN for thirteen years with mixed results. The Worcester station had a solid foundation of eighteen to twenty-four-year-old males but hadn't had much impact on "The Rock of Boston" with its lucrative core of twenty-five- to fifty-four-year-old male and female listeners. Sure, the two stations elbowed each other at times when competing for promotions with groups they shared on their playlists, but when it came to ad dollars, WBCN sat in a commanding position, even before the arrival of Howard Stern. However, with 'BCN's realignment away from album-oriented rock and toward "modern rock," shifting to a younger target demo of eighteen- to thirty-four-year olds, the presence of harder-rocking WAAF in the mix became undeniable. Then, the odds changed again in favor of the Worcester station as a new tag team of Opie and Anthony arrived from Long Island. The pair took over the afternoon slot at 'AAF and soon demonstrated their own low-brow path to "greatness" with stunts like "Whip 'Em Out Wednesdays," in which women were encouraged to expose their breasts to male drivers. The campaign became so popular in just a few weeks that the state police had to take a stand in response to those citing safety concerns over lapses of driver attention. To assuage their opponents, WAAF's managers slapped the pair's wrists by suspending them for two weeks. 'BCN's Shred reasoned, "We had

Stern, so we opened up the can of worms. There certainly wasn't any good taste involved; it was how far these jocks could push the limits."

"I listened to Howard Stern and actually enjoyed him," said Steve Strick, "but I was not a fan of Opie and Anthony. With Howard, there was some sophistication there and some professional writing going on; it wasn't just two goons making fun of boobs." But like or dislike their approach, the new morning team's brand of raw humor proved effective and invasive, finding fertile ground with 'AAF's existing audience and then growing beyond. Eileen McNamara writing in the *Boston Globe* labeled "O & A" as "two witless disc jockeys at a second-rate rock radio station in Worcester"; nevertheless, this supposedly obtuse team was soon significantly impacting the radio ratings in the Boston market.

"'AAF had always been respectful and never had the balls to actually go after the station in the way that Opie and Anthony did," Mark Parenteau commented. "They were the first ones that really saw that 'BCN could be vulnerable." O & A made a very clear public target of Parenteau, attacking his character relentlessly on the air. "We believed that rock and roll is sometimes stronger, if not the strongest, in afternoon drive," WAAF's vice president and general manager at the time Bruce Mittman pointed out. "In our research, we kept seeing the word 'old' in relation to 'BCN, so O & A attacked the station, and Parenteau, for being 'old and tired.'" Although a success for nearly two decades, 'BCN's afternoon jock was, admittedly, spending far less time living the rock and roll lifestyle than he had before: "The station was hugely successful; we were making tons of money. I lived in Hopkinton then, and I had a house in Vermont. I'd go in every day and do the show, but I didn't hang out on the scene like I had in the past."

"Mark became much more domestic, and there's nothing wrong with that," Oedipus added. "He grew up, got older, and had other responsibilities; but he lost that youthful quickness. He was still a funny bastard, but Mark was no longer relating to the younger audience."

Opie and Anthony took it one step further and began ridiculing the veteran DJ about his homosexuality. Parenteau explained, "They attacked me for being gay, and it was uncomfortable because Oedipus had always said, 'Don't talk about it on the air.' I'd mention things in double entendre, so if you were gay, you would know I was too, and if you weren't, it could be taken another way. Oedipus didn't think it would be good for business, nor did I, to be the gay disc jockey or the gay radio station."

Summer '97, goofing off with Snoop Dogg backstage at the River Rave. (Left to right) "Chachi" Loprete, Oedipus, Snoop, and Mark Parenteau. Parenteau's days at 'BCN are numbered; he will leave within six months. Photo by Leo Gozbekian.

"So, do you take the high road and not respond to anything, then hope it goes away?" Chachi Loprete asked. "Or, do you bring yourself down and fight them on their level? We did nothing, for a long time. Parenteau kept his mouth shut through the whole thing, because how do you fight that?"

Parenteau continued, "I wanted to go balls to the wall with them over it; but of course, I would have had to acknowledge that I was gay. [But], the way the gay thing has turned out for lots of celebrities since, I don't think it would have made any difference. In hindsight, I think it would have made great radio; it would have been no skin off my nose if everyone knew I was 'the guy.'"

"The fact was that Mark couldn't beat Opie and Anthony," Oedipus countered. "At that point, he had the talent, but he didn't have the anger. They were really nasty, and it was tough to play in their ballpark. That really changed the face of it."

"Their guerilla listeners would drive by the house, make noise, and throw cans in the driveway," Parenteau confessed. "Suddenly I felt I had no privacy in my life, and I ended up retreating; I didn't want to run into this 'AAF thing all the time."

In the winter 1997 Arbitron book, WBCN placed seventh in the market for all listeners twelve years and older, continuing to win the eighteen to thirty-four male ratings race by scoring an 11.6 to AAF's 7.9. Still, while Howard Stern could be credited with maintaining the comfortable margin, the afternoons were definitely losing ground to Opie and Anthony's unrelenting assault. By the summer ratings period six months later, Opie and Anthony soundly trounced the veteran in the eighteen to thirty-four male battleground by scoring a 13.3 percent audience share to his 9.4. Tensions between Oedipus and Parenteau, as they disagreed on a winning strategy, often boiled over, their clashing and flamboyant personalities reduced to shouting matches or steely episodes of silence. By the fall, Oedipus had concluded, "It was time for Mark to move on; he was no longer relating to the audience." Parenteau was not surprised but disappointed, believing that the main reason for the cold shoulder was actually a financial one: "Within the structure of CBS they had a lot of DJs that were making a lot of money; we're talking a quarter-million dollars a year, and in some cases even more than that. Mel was trying to pare down that cost and replace million-dollar talent with forty-thousand-dollar-a-year jobs, which he did. So, Oedipus was playing this whole Mel game at the time; he had to." In the face of declining ratings, Oedipus and Tony Berardini chose not to renew Parenteau's contract, which meant that the DJ was out of a job in early November '97. "There was this whole statement [drawn up] that said I wanted to leave because I wanted to go do other things." WBCN's press release stated, "The venerable disc-jockey plans to take a short break from his radio show to pursue several entertainment-related opportunities that were recently presented to him."

"That just wasn't the case," Parenteau refuted. "I didn't want it to go down that way."

Dean Johnson announced the startling news in the *Boston Herald*, also making clear that he didn't believe 'BCN's press release for a minute. "He's the last of the station's longtime jocks to go, the final member of a gang that once included Charles Laquidara, Matt Siegel and Ken Shelton. Here's the real deal: Parenteau got older (he's in his mid-40's), WBCN's audience got younger (its prime demographic is now men 18–34), and in these bottom-line times, he's just too expensive." The writer was also curious that the station hadn't announced a suitable goodbye celebration for the legendary jock: "Waiting for the big Parenteau farewell bash? Don't hold your breath. Rather than a loud, kissy-huggy farewell week, WBCN is opting for a low-profile finale. And anyone who's listened to Parenteau over the years knows that, given the choice, he never does anything quietly." Johnson proved to be a prophet as his prediction of a mischievous outcome came true the very same day his *Boston Herald* article hit the streets.

On that 5 November, the soon-to-be-unemployed disc jockey honored a promise to appear at the prestigious Achievement in Radio (AIR) Awards being presented at the Marriot Long Wharf in Boston. Tom Bergeron hosted the glittering affair while Mayor Thomas Menino attended to present WBZ-AM's morning anchor, Gary LaPierre, with a Lifetime Achievement Award. "But while Bergeron got a lot of laughs and LaPierre got a standing ovation, the real attraction in the crowd was DJ Mark Parenteau, who was just cut from WBCN-FM after 20 years at the station," Susan Bickelhaupt reported the next morning in the *Boston Globe*. "Parenteau showed he was a trouper and filled his commitment to present an award. And he managed to use his air time at the mike to get some barbs in, noting that, 'I feel like I have shaken baby syndrome . . . but hey, life goes on,' and that the CBS [Television] 'Welcome Home' slogan should be 'Welcome Homeless.'"

"It was just the right line at the right time," Parenteau chuckled. "Everybody woke up to headlines about me in the *Globe* and the *Herald*; it really embarrassed CBS, and Oedipus banned me from the station. I was quickly excommunicated and not allowed to do a final, farewell show."

Amid the daily drama of Parenteau's battle with Opie and Anthony, it barely registered that there were some other lineup changes that had occurred. In June '97, Matt Schaffer returned to WBCN to reassume his duties as host of the "Boston Sunday Review." "Sometimes you *can* go home again," he told the *Boston Globe* after Oedipus hired his good friend back on a part-time basis to host the public affairs show, now on Sundays 7:00–9:00 a.m. The same week Schaffer went back on the payroll, Bradley Jay announced he was leaving to pursue his quest of being a talk-radio jock, vacating the midday slot he had filled after Ken Shelton departed. Oedipus told the *Globe*, "He wants to do something different, he wants to be a hip Larry King." Just before leaving, the jock hosted an intimate gathering of 104 WBCN listeners in a Q & A session and performance with his hero David Bowie, live on the air at Fort Apache Studios in Cambridge. As Jay exited on this personal high note, Bill Abbate, who would also burn the midnight oil preparing for his weekly role during Patriots season, moved in.

"It was a long transition for me to get to middays; I would do that shift up until September 2001." Like his two predecessors, Abbate loved the show for the many opportunities it afforded him to interview artists: "Lenny Kravitz came in, multiple times; that was always fun. Green Day showed up to be guest jocks once, and at the end of that visit, Tre Cool lit up this huge joint, right in the studio where the guests would sit . . . underneath the giant vent for the air conditioning!" Abbate watched in amusement as the smoke sauntered slowly upwards to be whisked into the opening and channeled to some distant part of the station. "Then, wouldn't you know it, all the sales people suddenly showed up in the studio!"

Oedipus had another pair of big shoes to fill once Mark Parenteau had been bumped from his afternoon radio home after two decades. The program director found his replacement in-house with a jock he had hired the year before to stake out the night shift when Stern moved off tape delay and into the mornings. "We had all been trying out for that evening shift," Neal Robert remembered, "but I felt my chances were diminishing. My forte was being a music person: a guy who was not bigger than the songs, but a companion on your journey with the music. But at that point, Oedipus really wanted another shock-jock on at night; he wanted 'the talk.'" The program director hired the vociferous and outspoken Nik Carter, who had worked with Robert at WFNX for seven years before landing a gig on the morning show at "The Edge" (WDGE) in Providence. "I was very familiar with Nik from his days at WFNX," Oedipus related. "A natural talent, he was one of those rare individuals who spontaneously always had something to say that was both interesting and entertaining. Plus, he was a contemporary music aficionado, the future of WBCN."

"I came out of the 'alternative radio' culture, and even though 'BCN was now 'alternative,' it hadn't yet developed into what it said it was. Being the heritage station, [it] had jocks who had been there forever, but for want of a better term, they weren't living the 'lifestyle.' It was like you were listening to the greatest basketball players ever, and now they were being asked to play baseball." Oedipus encouraged Carter not to hold back but unloose his raucous and unfiltered style in a radio show dubbed "Nik at Night," upping the ante of Bradley Jay's previous "Sex Palace" and adding music to a Howard Stern–like attitude.

Once again, not everybody welcomed the new hire at 'BCN, nor the attitude he brought with him; Bob Mendelsohn, for example, thought Carter

Nik Carter (center bottom) poses with Green Day. Also pictured from WBCN (left to right) Carter Alan, John Reilly, and Steve Strick. Courtesy of WBCN.

was "not a mature talent" but "a smart-ass punk." The new jock, however, pointed out that he felt he was carrying on the station's longtime mission: "I grew up in Cambridge listening to 'BCN; you were going to hear Stevie Ray Vaughan and Don Henley, but you were also going to hear some weird new band. More importantly, it was a cultural beacon: that's where my friends and I heard that Bob Marley had died; we were just devastated. And it was a station with a conscience: whether it was Charles railing against Shell Oil for apartheid or whatever crusade it went on, you always felt that 'BCN was going to come down on the side of right." Influenced as he was by the station's past, the new DJ ran with the crowd of eighteen- to thirty-four-year-old males that 'BCN now had to win. Accepted as one of their own, he was a representative who played the music of their generation. What a lot of them didn't realize, at first anyway, was his skin color. Jim Sullivan wrote in the *Boston Globe* in November '97: "Carter is a black man swimming in an ocean populated by a lot of very white men. Most of the DJs and most of the bands in his world are Caucasian. Black DJs? They're over at the urban dance or contemporary hit stations."

When Oedipus transferred Carter out of nights to take over for Parenteau, it was talked about almost as much as Stern's ascension over Laquidara a year and a half earlier. After all, it was the first major change in the station's afternoon shift since 1978. Oedipus assessed, "The music was changing, the station was staying young; [we] needed an afternoon DJ who could relate to our audience, grow in the position, reestablish afternoons, and carry on the great tradition of unique WBCN air personalities." The stage was now set for an epic cage match. Opie and Anthony began grappling with Carter the very first day, as he remembered: "They called it 'Black Monday,' and they were calling kennels around the city looking for their little black poodle 'Nik.'" The WBCN disc jockey immediately returned the jabs, with the results making headlines in both Boston dailies. "Over at WAAF, Carter's competitors have been trying to identify Carter as black, not always subtly, by means of verbal volleys and website postings," the *Boston Globe* reported. "On the air, WAAF has called Carter 'the dark-skinned lover' and 'disco boy.'" The *Herald* mentioned that Oedipus had provided a cassette of an Opie and Anthony show: "Some of the tape's excerpts include the duo saying: 'He's a big brown turd that stinks . . . Disco douche . . . Lionel Ritchie's love child.'"

Back and forth the battle raged, like artillery barrages over the front lines: inflammatory comments about Carter's color and then returning allegations of racism from WBCN. "It was probably one of the filthiest battles in rock radio history," Carter observed. "I was stressed out of my mind and on Paxil as a result." As the conflict deepened, the *Herald* commented, "It's just plain ugly and, at the very least, downright stupid." But it continued, for months, with Tony Berardini and Oedipus dragged into the fray, accusing WAAF and defending their own, while their counterparts, Bruce Mittman and program director Dave Douglas, did the same. "It stands as the most reprehensible radio experience of my life and appalled me to the core of my being," Oedipus stated. "This was not art, this was not competition; this was out and out hatred."

"I never felt there was a racist attack," Mittman contradicted. "Corporately, no one in WAAF management would have supported that. No one cared if [Carter] was black, white, or orange." While believing that statement may or may not seem difficult, what is true is that O & A were never actually caught using the *n* word on the air. But the damage was done, nevertheless, as members of their audience pressed the attack. Carter related, "I'd pick up the phone and there would be some 'AAF listener there [saying,] 'Nigger, nigger; nigger, nigger . . . Opie and Anthony rule!' And I couldn't really get angry at all these kids because they didn't really realize what they were saying; they were desensitized."

"That's where Opie and Anthony and I differed," Mittman stated. "'You might not be saying things directly, but you're encouraging people to do things that you are responsible for'; that was a constant argument. Rock and roll is a bad-boy business but not an irresponsible business."

The end of the O & A/Carter melee came abruptly, six months later, when the WAAF afternoon team initiated a poorly conceived April Fool's Day stunt in which they announced that Boston mayor Thomas Menino had been killed in an automobile crash. "The 'joke' was not so funny in the homes of the mayor's distraught relatives who had to field condolence calls," the *Boston Globe* reported. When Menino applied intense pressure on the offending station by filing an official complaint with the FCC, the threat raced to the very top of the WAAF ownership hierarchy. "Steven Dodge, chairman of American Radio Systems, met with the mayor to apologize for his disc jockey's lack of 'basic human decency,'" the *Globe* continued. "Their stunt could not have come at a worse time for American Radio Systems. The Boston-based broadcasting company needs FCC approval to be acquired by CBS Corp." How ironic that during the entire O & A versus Carter bloodbath, the home team's parent corporation had been negotiating to buy the company that owned WAAF. For a cool $2.6 billion, Mel Karmazin would scoop up American Radio Systems, although the Justice Department required that its four Boston stations, including 'AAF, be divested. Meanwhile, Menino was no dummy; by refusing to back down, accept an apology, and withdraw his complaint, WAAF had no choice but to fire Opie and Anthony to avoid the possibility of being stripped of its license and possibly derail the gigantic business deal on the table. The infamous tag team moved on; but in another irony, CBS turned right around and hired O & A to go on the air two months later at WNEW-FM in New York City.

Life didn't get much easier for the 'BCN jocks, despite the absence of WAAF's stars. The battle orders had already gone out and were received by the station's eager audience, many of whom gleefully carried on the mission of subverting "The Rock of Boston" whenever they got the chance. "I had no interactions with the 'AAF jocks," DJ Melissa related, "but I did have interactions with their fans. And to give 'AAF credit, they did a good job brainwashing those people." Bill Abbate mentioned, "It was no longer a matter of their jocks targeting us; they figured out a way to get their audience to do

Scott Weiland at the WBCN River Rave, 1998, with (left to right) Juanita, Nik Carter, Oedipus, Steve Strick (top), Bradley Jay, and Melissa Teper. Courtesy of WBCN.

it. So, there was this whole stretch of time where, depending on the band, you knew that there was going to be some riled-up 'AAF fan who was going to do something stupid." In July '98, Juanita received the assignment to do a solo live broadcast from "Ozz-fest" at the Tweeter Center. "Exciting day but difficult," she remembered. "I did interview a lot of people, like Lemmy [from Motorhead], the guys in Korn, Fred Durst of Limp Bizkit; but it was very much an 'AAF crowd at the show. There was this point where we were up on the lawn walking; I had a microphone and the engineers had bags with the 'BCN logos on them. People were not happy to see us up there; they started throwing stuff at us. Because there was so much hatred instilled by WAAF toward 'BCN people, I actually felt like we were in a dangerous situation, [so] we just got out of there as fast as we could."

WBCN wasn't merely a sitting duck, though; its staff learned to expect some animosity at certain harder rock shows. Mark Calandrello, known as "Cali" to his coworkers, who arrived in early 1998 and became Chachi's full-time assistant, signed on just as the Opie and Anthony assault reached its peak.

The racial remarks really kicked the rivalry into high gear. No matter what it was, if it fit the audience, we were battling with 'AAF over everything! I was on the WBCN "Street Team" at that point, and we brought in a bunch of college kids who weren't afraid of a fight or getting their hands dirty. We would terrorize the other stations, especially 'AAF, because we were sticking up for Nik. We'd go into their backyard at the Worcester Centrum, where they'd be broadcasting from a bar next door. We'd wait outside for them to go live on the air and run inside with a megaphone, shouting "Rocko ['AAF night jock], you fat fuck! You suck!" When they would leave their van at concerts on Lansdowne Street, we would wrap them in 'BCN roller banner and put 'BCN stickers all over them; then we'd take pictures and put them on our website.

Despite the bad blood, though, Cali didn't believe the intense competition was necessarily a bad thing: "We had a hard-core rivalry that lasted for years; other than that Opie and Anthony racial stuff, it made for great radio." Derek Diedricksen, known on air as "Deek," joined the WBCN DJ lineup in 2000: "We did a broadcast from a Korn show or something, and they had the radio stations, 'FNX, 'AAF, and 'BCN, all together in this tiny room. Everyone was sitting there being awkward and stupid, *not* talking to each other, like this was some religion by which we must abide! So, I went over and said hi to Jay Ferrara, who was the 'AAF guy up against me [on the air], and he was nice enough. Two days later, he was back at it, talking crap about me on the air. But, hey, it's entertainment; it's a job. I never took any of it seriously."

In June '99 the competitive spirit between the two stations became national news when Limp Bizkit organized an unauthorized show in Boston to promote its forthcoming new album *Significant Other*. "They were the biggest band at that time; we were playing Limp Bizkit every couple hours," Juanita recalled. But WAAF had been playing the metal/rap outfit since the very beginning, and the station had an entrenched relationship with the group and its record label. "I was on the air and we knew the show was probably going to be that day, but the band wasn't going to release the location except on 'AAF." Limp Bizkit's plans to do a rooftop concert, like the historic Beatles ploy in 1969, needed to be clandestine; a surprise attack might buy the group enough time to do a short set before the police moved in to shut it down. Juanita continued, "Shred just happened to be walking through the Fenway, looked up at one of the buildings, and noticed there was a band setting up on top of one. You don't see that too often! He called and told me. 'Omigod!' I said. 'It's the Limp Bizkit show!'"

Juanita didn't hesitate; she shared her precious nugget on the air within moments. "I gave out the location before Limp Bizkit could call it into 'AAF. As it turned out, MTV [News] was driving in the car with the band [filming for a documentary], listening to 'BCN, and they heard me announce the secret location." A video of that moment ran on MTV, and in it, Fred Durst (the band's lead singer) commented, "They got a war going on here," before hustling to the garage roof right around the corner from WBCN's studios at Fenway Park. MTV News reported, "The crowd of about 1,500 fans discovered the show's secret locale through a local radio station; so did the Boston Police, who shut down the illegal concert after a five-song, 20-minute set." Obviously, Limp Bizkit and WAAF were not pleased, but over at 1265 Boylston Street, the mood was ecstatic. If WBCN ran itself like the military, Oedipus would have held a ceremony and pinned Silver Stars on both Juanita and Shred as members of the staff looked on respectfully, doing their best impressions of standing at attention.

WBCN's personality began to shift as the new-music epicenter polarized to a harder scene. Playing a wide range of alternative music in '94—everything from the intensity of Jane's Addiction to the thoughtful musings of Tori Amos—the overall sound of the station had changed as 1999 arrived, the focus now lingering on a fresh crop of bands playing heavier rock and rap music. To the radio industry, although WBCN was classified as a "modern" station and 'AAF as "active rock," the number of artists that had become common to both formats sharply increased, often making the two stations virtually indistinguishable. "When I got there," Juanita remembered, "'BCN had just started playing Green Day and it was moving to a lighter alternative sound: stuff like the Verve Pipe and a lot of female artists [such as] Sneaker Pimps, Portishead . . . Oedipus's 'Women in Pain,'" she laughed. "It definitely took a late nineties left turn into that whole nu-metal, rap-rock thing." WBCN chased the male eighteen to thirty-four ratings into the heavy-metal rabbit hole, and the playlist now caught up with the edgier tone already in place on the rest of the station. While Howard Stern ruled the mornings with his coarse humor and coterie of impertinent pranksters, Nik Carter held much the same sway in afternoon drive, and (stereo) grunts from the gridiron ruled NFL Sundays. It was abundantly clear that WBCN's emphasis, in all ways, centered squarely on a battle of (and for) testosterone.

In 2000, the WBCN River Rave had swelled to colossal proportions, moving fifty thousand tickets to sell out Foxboro Stadium. The top-shelf bands

Juanita blew the whistle on the secret WAAF show
with Limp Bizkit. Photo by Andrew King.

performing included Stone Temple Pilots, Filter, Cypress Hill, Powerman 5000, and local heroes Mighty Mighty Bosstones and Godsmack. The daylong festival included onstage appearances by all the 'BCN jocks; a B stage (where Juanita's band Heidi performed, among others); a Rave tent in which pulse-pounding techno music reigned nonstop (featuring DJ Bradley Jay, who had returned to 'BCN as a part-timer); an electronic gaming area; and a "cyberpit" with Internet options. "The Raves were great concerts, 'Chachi' Loprete observed, "a million moving parts." The show stood as a towering triumph; five years after the first humble concert (the one actually by a river), the event had grown to fill the largest outdoor entertainment venue in New England. That achievement was overshadowed, for better or worse, by a record 126 arrests and what Dean Johnson labeled in the *Herald* as "breastapalooza." During the show, women were encouraged to flash their assets, the images relayed and enlarged on the gigantic onstage screens for all to see. Johnson wrote, "The video expo went on for nearly the entire 12-hour concert. At one point, a 'best of' segment was shown in slow motion." All this liberated activity raised the ire of local police and Foxboro town officials, who voiced their displeasure. Balancing their delight over the bad-boy image of the Rave with a need to assert some damage control over the official reaction, WBCN officials issued a rare apology, blaming the video production company for the displays. Dean Johnson smirked in print, "Considering WBCN's bawdy programming—and the fact that the station's disc jockeys joked about it onstage and on the air, the apology is more than a little disingenuous."

In a surprising development, considering just how much vitriol had been spilled out on the airwaves between Nik Carter and his WAAF tormentors just two years earlier, Opie and Anthony appeared on the massive River Rave stage, wireless microphones in hand, to motivate the WBCN audience. "The two weren't off the Boston airwaves for long," the *Boston Globe* reported a month later. "The difference is that now one of Opie and Anthony's prerecorded spots invites listeners to make an obscene gesture if they see the WAAF van, because these days, the duo are regulars on WBCN, phoning in their 'bits' from their New York base at CBS sister station WNEW." The *Globe* article mentioned that the tag team also made "caustic appearances on WBCN's afternoon show, as guests of the similarly puerile Carter." To any observer, this represented a vast change of heart.

"People say, 'How could you forgive them?'" Carter reflected. "Opie and I bumped into each other in New York City and talked. Then they [both] came to Boston on a Saturday afternoon and we drank, and laughed, and fought, and yelled at each other at a bar for like five hours. They said, 'We went too far'; 'that's not who we are'; 'we were wrong.' If they were big enough to apologize, I was big enough to say, 'All right, I got ya.'" That WBCN had learned how to scrap it out on WAAF's level was quite obvious after the 2000 River Rave. "The battle woke Oedipus up and made him competitive," Bruce Mittman observed from across the trenches. "It also forced us to be on our game, all the time. If you don't own the streets in rock and roll, you're not a rock and roll station." The *Boston Globe* commented, "The two stations have nearly switched places, with WBCN abandoning its earlier ethos to co-opt WAAF's frat-boy sound and attitude."

Across town in August 2000, Charles Laquidara ended his run at WZLX to move on to retirement in Hawaii. As Boston said goodbye to their legend, the distance between where his former radio home had been and where it was at the moment lay heavy in the minds of many former listeners. This was surely not your father's 'BCN; aloha to that. Surprisingly, though it now reflected the tastes of those aggressive, disaffected, often sophomoric males it was trying to attract, 'BCN's grittier image still managed to generate a windfall at the bank. The *Boston Herald* peeled back the image in May 2000: "WBCN is a master of illusion. It pretends to be a rebellious upstart and enjoys massive street credibility. But in reality it's a slick, carefully programmed machine that enjoys impressive ratings and sells ad rates that are the envy of corporate radio." The article revealed that only sister station WBZ-AM grossed more revenue locally, with 'BCN's take in 1998 (the most recent numbers available at the time) at a whopping $29.2 million. "What a remarkable paradox. The station that plays cutting-edge rock with over-the-top, potty-mouthed personalities rakes in more cash than any other music signal." This occurred despite the fact that, as the *Herald* also mentioned, WAAF beat WBCN with men eighteen to thirty-four in every day part except Howard Stern in the winter 2000 Arbitron ratings. "We kept growing the revenue, and the challenge was to continue finding ways to do that," Tony Berardini acknowledged. The obvious dividend on 'BCN was still Howard Stern, whose ratings were, in a word, untouchable. "At one point," the general manager continued in amazement, "we were selling [select] spots on his show for $25,000 apiece. We were getting television rates; that was just insane."

Even though trash-talking, shock-jocking rants from Howard Stern, Nik Carter, and Opie and Anthony now dominated WBCN's image, the beating heart of the station was still the music. Juanita remembered one of her favorite moments at the station, which occurred in July 1997:

> Joey Ramone was in the studio, guest DJ for "Nocturnal Emissions" on a Sunday night, and Albert O was operating the board for him." I came in to do the local music show, "Boston Emissions," right after that. Albert said, "It might go a little late." I said, "We better check with Oedipus," so I called him. He told me, "Let Joey go for as long as he wants."
> "Uh, okay. What if it's a really long time?"
> "He can play records for as long as he wants."
> "All right." So, Albert left, and it was just Joey Ramone and me in the studio; he was in the guest chair, and I was behind the board cueing up his albums. I had a local band that was coming in for an interview on "Boston Emissions"; they got there and were just, "Omigod! It's Joey Ramone." So Joey ended up interviewing the band with me on the radio. It was surreal! After that, I said, "Do you want to keep playing music?" He said, "Yeah!" We went on like that for five hours . . . until three in the morning!

Adding his time with Juanita to the two-hour portion with Albert O, the lead singer of the Ramones had put in nearly a full workday on the air. But the privilege of "hijacking" the station had been extended out of loyalty.

Considering how long Joey had known Oedipus and had been coming up to 'BCN from New York, he was thought of as part of the radio station's family. So, in this visit, which would be just a handful of years before his death from lymphoma in 2001, Joey Ramone didn't show up out of record company obligation or mandate; he was just having fun, "working the counter" at a "family business" that he loved.

Moments like these, when a 'BCN DJ propped open a creative window for an artist, letting convention slide despite the rules or the logic of commerce, had been one of the essential ingredients of WBCN since the beginning. Respecting the artist and the music was always part of the mission.

Success bought freedom, but this "family business" couldn't stave off the outside world forever, especially when challenged by its own company. In September 1999, Mel Karmazin, CEO of Infinity and CBS, announced that the already-enormous corporation would be merging with another media giant, Viacom, under the firm (some said, ruthless), guiding hand of seventy-six-year-old Sumner Redstone. Karmazin first proposed buying Viacom, but when Redstone refused, he suggested that Viacom acquire CBS in a complicated merger arrangement. By the time the deal was finalized, then approved by Washington, the arrangement cost Viacom $39.8 billion, one of the largest media transactions on record. The *New York Times* reported that the merger would create a new company that "makes, distributes and broadcasts television programming, blankets the nation with radio stations and billboard advertising and owns and operates amusement parks." With Karmazin and Redstone as co-CEOs of Viacom, and Karmazin retaining his status as president of Infinity, the media conglomerate now included, according to the *Times*, Infinity's 160 radio stations, the CBS television network, CBS Sports, Internet properties like cbs.com and marketwatch.com, the country music networks CMT and TNN, MTV, Nickelodeon, Showtime, Comedy Central, Sundance Channel, Paramount Studios, five regional amusement parks through Paramount Parks, 6,400 Blockbuster stores, the Simon & Schuster publishing giant, and billboard advertising through TDI Worldwide and Outdoor Systems. Clearly, this enormous corporate galaxy was light years removed from the tiny world in which Ray Riepen had launched his rock and roll experiment back in 1968. WBCN, with its questing rabble of radio pioneers who had once vowed to "Kill Ugly Radio" in Boston (and did, for decades), was now just another point of light in Viacom's vast Milky Way.

SHINE ON YOU CRAZY DIAMOND

Opie and Anthony's brand of offensive radio gumbo thrived in New York City, proving that mediocre taste amongst radio listeners wasn't necessarily exclusive to the Boston market. Because of this success, and the previous gold that Infinity had mined by syndicating Howard Stern's show into multiple markets, CBS pursued a similar deal for O & A's afternoon show. Tony Berardini and Oedipus were now presented with a situation quite similar to the one that had occurred when Stern bumped Charles Laquidara off his longtime morning perch. If they didn't pick up Opie and Anthony's option in Boston, the offer would be made to someone else in the city. "And who was the station going to be?" Berardini posed. "WAAF; the station they started off with." Of course, everyone in the loop was all too aware of what had happened when the ignoble tag team worked for that competitor three years earlier. Now faced with another catch-22 decision, Oedipus wrestled with his conscience and his duty. "It was awful," he would say ten years later. "That was the beginning of the end. Radio had ceased to be what I liked, and what I envisioned it to be about in the first place." Nevertheless, the program director added O & A's show onto the 'BCN airwaves. Oedipus conceded that he did what he had to do to guard what was left at the station: "It was a collective decision, and I agreed to it as a way to protect my staff. The station was so revenue oriented by then that it was a matter of maintaining jobs." Yes, the station was certainly revenue oriented: *Boston* magazine reported that WBCN's take in 2000 was up to $38 million, only bested in the market by WBZ-AM.

"Now, all of a sudden," Berardini pointed out, "we had the two most important day parts on the radio station: morning and afternoon drive, as talk [shows] with Howard Stern and Opie and Anthony; yet we still wanted to be [known as] a music station!"

"With two talk shows and the Patriots, our music image was smoke and mirrors," Oedipus admitted. His second-in-command, Steve Strick, said, "I had to justify to the record labels why we were still relevant, and I had to do it every day. When you deemphasize the music, it has a huge effect; your clout with the labels disappears." O & A's arrival also disrupted the entire DJ lineup, bouncing Nik Carter to the midday slot, Bill Abbate to weekday overnights, and Albert O out of his full-time shift. Despite all that turmoil and the "talk versus music" fallout, O & A's show did deliver substantial numbers, ranking number 1 among men eighteen to fifty-four in its first ratings period.

But Opie and Anthony's simulcast show from Manhattan thrived in the afternoon slot at WBCN for little more than a year. The witless duo crossed the line again in August 2002 by broadcasting a segment in which a couple attempted sex in St. Patrick's Cathedral, mere feet from worshippers observing the Holy Feast of the Assumption, initiating a massive public outcry and outrage in the Catholic Church. When the religious hierarchy brought the full weight of its political clout to bear, not only Infinity, but also the entire CBS/Viacom parent, suddenly became vulnerable. The company quickly pulled O & A from the airwaves, an action that appeared to appease the church but not the FCC, which slapped Viacom with massive fines for indecent radio content. While dealing with these legal issues, the company, worried that the offending pair could be hired away to work for any competition if they were fired, paid O & A the balance of their estimated $30 million contract to sit at home for two years and watch

TV reruns. To many hardworking employees of Infinity, CBS Radio, and Viacom, this was seen as the ultimate insult as Opie and Anthony laughed all the way to the bank.

Oedipus went back to 'BCN's basic playbook, returning music to the mix. Nik Carter returned to the afternoon shift, retaining a good deal of the irreverence its former tenants had spewed but also restoring music to its prominent role. 'BCN also embarked on a nationwide hunt for a new midday DJ by holding live on-air auditions. Adam Chapman, known to his listeners as "Adam 12," became one of the sixty applicants who traveled to 'BCN for their own tryout. He had begun his professional career as a part-timer at WFNX before leaving the market in 1999 to work full time for Clear Channel in New Mexico, but Chapman relished the idea of coming home to Boston. "Each [prospective] DJ had a sound clip on the website so people could listen to them and vote," Adam 12 pointed out; "it was very interactive and way ahead of things at the time." When the people's choice turned out to be the former 'FNX jock, Oedipus concurred and ratified the vote. "I think the reason he hired me," Adam 12 laughed, "was that I was the only person he didn't have to train to say Gloucester or Worcester correctly!" WBCN's newest DJ (and, incidentally, the last one Oedipus would hire) began his first official midday show on 24 February 2003. "My mom told me a story of when she was in high school. She would come home from school every day and tune in WRKO with Dale Dorman and all those Top 40 guys; then a friend turned her onto WBCN. Now, here I was working for that same station!"

"There was a fraternal irreverence," Adam 12 recalled of an institution that was about to turn thirty-five years old as a rock station. "I would see Oedipus riding up and down the hallways on his Razor scooter, talking to record reps through the earpiece on his phone!"

"When Adam 12 was on 'FNX, he taped me giggling on the air at 'BCN," DJ Melissa remembered (she was renowned for her infectious tittering). He looped it on a tape and used it as a sound bed when he would speak at the mike. Then he became the midday DJ at 'BCN, and we had never met." After Adam 12 encountered Melissa, politely offering up a hello and a compliment, the red-headed veteran purred back, "'I didn't know you were such a fan. If you want, you can turn on the mike and I'll just giggle for you!' He looked back, shocked that I remembered. 'Oops!' he said, all embarrassed. I walked away, 'My work is done!'" she laughed. DJ Melissa would be part

Figure 16.1 Adam 12 comes to WBCN in 2003. With WBCN going out of business in 2009, then WFNX sinking in 2012, this unlucky Boston jock has had, not one, but two horses shot out from under him. Photo by Janet King.

of the final puzzle piece to be rearranged in the post–O & A lineup as she finally won her full-time shift. Melissa Teper discovered an on-air chemistry with part-timer Deek (Derek Diedricksen) that impressed Oedipus, so he organized the two into an official team and gave them 'BCN's evening shift in 2003.

"I escorted winners to the 2002 Super Bowl," Deek remembered. "We flew down to New Orleans and back in one day! We also went to the Playboy Mansion for a broadcast [promoting] Pete's Wicked Ale. It was a pointless pissing away of money for them, but a lot of fun for us."

"I was on that Playboy Mansion trip," mentioned Bill Bracken, chief engineer at WBCN for six years, "but the real event was hiking with Deek up to the 'H-O-L-L-Y-W-O-O-D' sign to hang a 'BCN banner! It was silly, because the letters were forty-feet high and our little banner wouldn't even get noticed. We cut through people's yards, didn't know it was a national monument, and at some point we tripped an alarm because a voice came out of a hidden speaker saying something like, 'You will be prosecuted if you proceed!' We panicked and ran back down the hill!"

"I got sent to interview Marilyn Manson in Lowell and he was huge at that point," recalled Juanita. "I went with Bill Bracken and we were taken into this backstage area with black curtains, lit only by candles. Marilyn Manson was sitting there, all in white, with dark sunglasses on that he obviously didn't need. I was terrified. But I gave him this present: an eight-foot-long rosary carved out of big chunks of wood. I hoped it would break the ice, and he loved it! We talked for forty, forty-five minutes . . . a long time."

"Coldplay played the River Rave," DJ Melissa recalled. "They were huge in England but not known at all over here. The crowd was there to see Marilyn Manson and they were throwing shit at Coldplay onstage. Bill

Abbate and I had to interview Chris Martin right after he got off and we were just trying to joke with him, laugh the whole thing off, but he was having no part of it. He was taking himself way too seriously. So, ha, ha, I suggested he go see a therapist . . ." She let the last word hang in the air for a couple seconds and then added sheepishly, "It was very awkward."

"I interviewed the Cure," Nik Carter remarked. "At one point, Robert Smith was drinking a Sam Adams Cherry Wheat, and I said to him, 'Hey Smitty? Where's mine?' And he just handed me the bottle. I looked at the bottle with his red lipstick all over it and I drank it down. That was the closest thing to a gay experience I ever had!"

In July 2002, Infinity promoted Oedipus to vice president of alternative programming, overseeing WBCN and fifteen other top-market radio stations. However, the excitement of the appointment didn't last. "I was not allowed to take risks; everything had to be vetted by corporate. Cutting costs and raising revenues was the mantra." It's no wonder that there was an increased degree of anxiety in the CBS/Viacom boardroom. With the worldwide recession of the early 2000s eating into profits for most businesses, companies cut where they could. Subsequently, the amount spent on radio advertising dwindled, and CBS, like all radio companies, hit hard times. Tony Berardini said, "In 2001, after 9/11, everything changed when the revenue in the marketplace was so drastically reduced." Then it became worse: in 2002 the cost of actually doing business increased. "The Enron scandal sparked the Sarbanes-Oxley Act. What [the feds] were trying to do when that legislation was adopted was provide more transparency for public corporations, but what it meant was that you were constantly audited. Now, you had to provide a trail to track every one of your expenses, and all your revenue had to be accounted for meticulously." The new law required increased accounting attention, which meant more accountants, sometimes to the detriment of other departments, which had their staffs reduced. Berardini continued, "I remember going up to CBS Radio Corporate, which was in the Viacom Building in New York; I walked out onto the floor and there were twenty or thirty offices and, perhaps, forty cubicles there. Of those, probably two or three were devoted to programming and a few more devoted to sales. The rest of it was all accounting people! Symbolically, it was indicative of how radio was changing at that point."

Meanwhile, enormous changes in technology began to affect how people bought or listened to their music and had a major impact, not only on WBCN, but also on every radio station in the country. "It was the era of the Internet onslaught, Napster, and downloading," Nik Carter said. "I remember thinking that what happens between the songs is even more important, because now the listeners could get the music even before we did!"

"We wanted to be the new-music station but get an audience too, and it was tough," Steve Strick added. "The music was becoming portable and customizable to everybody [through] the Internet and the iPod; that had a huge impact on how people used radio. We were never able to get the gangbuster ratings we'd had, maybe, eight years or so previous to that."

"You had the radio competition taking a chunk, and now you had the 'net' and iPods coming in," Berardini said. "That was coupled with the station's need to try and keep big ratings, to keep growing the revenue. We were casting our net wider, but it wasn't there; it had been chipped away."

"'AAF and 'FNX had both managed to carve out niches for themselves, [but] there wasn't that same audience loyalty or energy revolving around our radio station as before," Strick said. "It was difficult to keep generating the interest in what we were doing musically. The younger audience that we had hoped to bring up like we had always done previously was abandoning us in droves."

By 2004, Tony Berardini had asked to leave his position leading the efforts of CBS Radio/Infinity's Boston radio cluster, particularly his role as general manager of WBCN and WZLX: he had had enough. The company, loath to lose his talents, however, retained the veteran to work in talent acquisition. No longer would Berardini have hands-on control of WBCN's course, as he had for over twenty years. Now, Mark Hannon became the new general manager of the station, along with WZLX and WBMX. Meanwhile, in New York, despite achieving one of the most enviable positions in media as chief operating officer of Viacom, Mel Karmazin had become miserable after years of friction with the co-managing Viacom CEO, Sumner Redstone. With Karmazin's departure in June '04, rising as the CEO of Sirius Satellite Radio barely five months later, CBS lost its chief advocate for radio in the entire Viacom hierarchy. In WBCN's first-floor office at 1265 Boylston Street, where Elvis Presley's handsome ceramic face still gazed out at the Red Sox fans and the street urchins who ambled by, at least one member of that hierarchy grieved that his boss was moving on. "When Mel left the company," Oedipus reflected, "it was obvious to me that my tenure was coming to an end. The new powers no longer understood what 'BCN stood for; they stifled individuality. Creativity became a carefully defined set of rules." The porcelain bust took in its last impressions of Back Bay street life before the movers arrived with their boxes and packing tape; Elvis was about to leave the building.

On 4 June 2004, both the *Boston Globe* and the *Herald* announced the inevitable, yet still astounding, news: Oedipus was stepping down after nearly thirty years at WBCN, the last twenty-three of them as program director. He would retain his VP stripes and continue to work for the Infinity/CBS national staff, while staying on at WBCN until a replacement arrived. Dean Johnson at the *Herald* interviewed Paul Heine, editor of the radio trade journal *FMQB*, who told him, "It's the end of an era. Oedipus played a pivotal role in bringing punk rock and new wave to the airwaves. He's one of the most colorful, controversial, outspoken, and influential program directors in the history of rock radio."

"I was downsized like so many others, but at least I was able to 'retire' from WBCN," Oedipus later commented. "They asked me to continue my corporate position as VP of alternative programming, ostensibly out of respect for my years building the company, [but] my input was too divisive. When I was finally released, I had become just an irritating number on a ledger."

"Oedipus left, and it sucked," Juanita recalled. "I think we all felt it was coming because he wanted to do things that he wasn't really able to do, and he didn't like going through all those corporate channels. He lost the love of it."

"When he left, it was like somebody popped a balloon for a lot of us, certainly for me. A lot of the spirit was gone," said Mark Hamilton, who had been hired by Oedipus four years earlier. The weekend, and soon-to-be evening, jock had come from WFNX and immediately bonded with his boss after telling him that, as a kid, he had listened to and taped the program director's memorable night with Bradley Jay at the David Bowie concert in Sullivan Stadium back in 1983. "We were in his office. Oedipus said, 'Check this out!' He rifled through his file cabinet, found an envelope, pulled out a yellow napkin, unfolded it, and *the* cigarette butts were in the napkin.'" Hamilton stared at the seventeen-year-old Marlboro infused with rock star molecules. As a Bowie fanatic, he could appreciate its significance, even if most would think it was just plain weird. Hamilton lamented the departure of the one who hired him: "We all had a loyalty to Oedipus; he had taken care of his air staff, he always had your back."

Steve Strick, in his capacity as assistant program director, became the interim person in charge. Groomed by his former boss to be the successor, Strick hoped he'd be able to slide smoothly into the corner office and use the experience and skills acquired during his twenty-three years at WBCN to solve some of the station's issues and usher it into a new era. But this was not to be. "When Oedipus announced his retirement, instead of me being offered the job, I was told that I was welcome to apply for it like everybody else." The press release announcing the new hire came just five weeks later: Dave Wellington, the program director for Infinity-owned KXTE-FM in Las Vegas, would take over at WBCN on 2 August. "The day they made the decision, Mark Hannon took me to Starbucks around the corner and told me the bad news," Strick recounted. "That was the most devastating thing to hear in my entire career. But he offered me a lot of money to stay around and be APD/MD [assistant program director/music director]; I had nowhere to go at that point, so I accepted the offer." Hannon was particularly worried about the stampede of listeners away from WBCN after every Howard Stern show and how the station could recycle at least some of them back to its other programming. In Las Vegas, Wellington had been quite successful at solving this problem, as Rob Poole (his on-air handle being "Hardy"), who worked with the program director for six years and would soon be hired at 'BCN, explained: "KXTE was a flamethrower of an alternative radio station. We played what we wanted, we had Stern on in the mornings, and we had monster ratings; I think we were only number 2 to a country station."

"Who was this person coming in?" Mark Hamilton wondered. "People in Boston don't like a lot of change. Who was this guy to come in and tell us how to run 'The Rock of Boston'? Has he even been to Boston? Everybody was deflated." When Wellington moved into 1265 Boylston Street, one of his first acts was to connect with Strick, who remembered, "He took me out for a beer and we worked together pretty well, but it wasn't the same. He didn't know the market; that's why they wanted me to stick around. But I was powerless, and I realized pretty quickly that he was powerless too. They were calling the shots from New York."

"He had big shoes to fill, and I kind of felt sorry for him," Juanita mused. "But he did cancel 'Nocturnal Emissions,' and we were really bummed about that. His reasoning was, 'Why are you doing two hours of music program-

The Oedipus dynasty comes to an end. The program
director (on the left) declares peace with Dave Grohl.
Courtesy of wbcn.

ming that has nothing to do with the rest of the week?' Uh, 'cause it's cool?
But, I'll always love him because he eventually gave me a full-time shift."

"I know when somebody comes in, they want to put their own stamp
on things," Hamilton said, "so the new-music show went, then the River
Rave, then the Christmas Rave."

Strick added, "The brand name of wbcn, the history of the station, didn't
seem to be important anymore."

wbcn had already stepped away from heavily playing the "nu metal" of
Korn and Limp Bizkit, adding some classic material like Guns N' Roses
and Led Zeppelin back in the mix. With Wellington's arrival, that process
continued; plus, the era of encouraging djs to perfect their fast-talking,
attitude-driven raps between songs formally ended. Now the on-air approach seesawed the other way, as Hamilton confirmed: "The station
became music intensive again, to the extreme of almost no talk. At one
point we were doing two fifteen-second and one thirty-second break per

WBCN ends the River Rave series and begins "Band Camp."
New arrival Hardy chats up Perry Farrell with engineer Bill
Bracken in the background. Courtesy of WBCN.

hour." With 'BCN adopting Wellington's "less is more" strategy, the odd person out (other than the untouchable Howard Stern) was the personality-intense afternoon guy, Nik Carter. "There was a new program director and a new general manager; plus, I was making a healthy salary for the time," said Carter. "I knew my days were numbered." In January 2005, CBS Radio notified the jock that the company would not be renewing his contract. The split, though disappointing to Carter, worked out cordially enough; he used company connections to find his next job elsewhere under the CBS umbrella. Some of the other 'BCN jocks would also move on at this time: DJ Melissa departed by choice in November '04 for a radio job in LA, and her former partner in crime, Deek, was ousted just six months later.

In April 2005, CBS/Viacom decided to consolidate the physical location of its Boston radio properties. WBCN left its home of twenty-five years in the "crime-free Fenway" and moved to 83 Leo Birmingham Parkway in Brighton, where sister station WODS-FM had already been based for a few years. "The 'BCN move was pretty stressful," acknowledged Bill Bracken, who masterminded the operation. "I lost fifteen pounds during the process." Actual construction of the spacious air studio (the former site of the building's boiler) and three production studios in the basement took four months, and the usable space for the entire station doubled to fourteen thousand square feet. The move allowed Bracken to upgrade 'BCN's eighties-era equipment and also uncovered bits of the station's past as the Fenway location was dismantled. Deek, whose final days at the station were counting down rapidly by this point, said, "When they cleared out 'BCN on Boylston Street, there was this beautiful Audubon book on birds that was being thrown out. It said, 'Property of Charles Laquidara' in it. [I] never met him, actually, but I've had his book on my shelf ever since." Reaction to WBCN's new home was mixed. "The new studios were like walking into a dentist office: no bumper stickers on the walls and a notice not to hang stuff up on the walls," Deek complained. "The old place had been kind of like an 'Our Gang' of rock; it was beaten, worn in, and threadbare. If you put a black light on that couch in there, you didn't want to know what you'd find! That's what a rock station should be."

"It was a very sterile environment," Juanita concurred; then she brightened and remembered, "but it was great to finally find a place to park during Red Sox games!"

It's easy to blame the immediate people in charge, Dave Wellington and Mark Hannon, for 'BCN's downward spiral after the move. But clearly the station had lost its bearings and wandered off course years earlier. Tony Berardini looked back all the way to 1996 for a signpost: "Stern going on in the mornings was the point of no return. One decision led us down the path that led to all the other ones. We kept going down that road [and it] led to the demise of 'BCN eventually." The arrival of Stern forever changed the complexion of WBCN, and the swapping of the shock-jock with Charles Laquidara marked an abandonment of the station's character for many a longtime listener. After that point, there were always two 'BCNs: the rude and often-insulting talk-fest version and the musical side that connected directly to the station's legacy. As fractured a diamond as it was, 'BCN still made money for the company, but the clock was ticking.

Soon after Wellington arrived, Stern announced he was abandoning the terrestrial airwaves to follow his former boss, Mel Karmazin, to Sirius. In October '04, the shock-jock revealed that his final show for CBS would be in December 2005. "So, then, Howard Stern was on the air for more than a year," Hardy spat, "prompting people, every day, to switch over to satellite radio because that's where he was going!" This disheartening unfaithfulness to his CBS coworkers was effective: after the Stern show became a monthslong commercial for his future radio home, the number of Sirius subscribers rocketed from 600,000 to 2.2 million.

"When Howard left and went to Sirius," Steve Strick said, "we lost one of the most compelling reasons for people to tune in WBCN. Those listeners were gone." That story played out on CBS affiliates coast to coast, all of whom would experience a loss of ratings and revenue as Stern's estimated twelve million daily listeners either struggled to find new morning entertainment or shelled out $12.95 per month to follow him to satellite radio.

CBS weighed its options during Stern's protracted fourteen-month goodbye, and one of them led to former Van Halen lead singer David Lee Roth. With his performing career, at the time, in the rearview mirror, Roth entertained the idea of being a syndicated radio host. He had the rapid-fire wit and reputation necessary, so serious talks with CBS advanced rapidly. The potential radio hire was slotted in to do an on-air audition, which happened at sister station WZLX in March 2005 during a week of morning shows. Chachi Loprete, now promotions director for both WBCN and WZLX, struggled with Roth during the weeklong audition:

> I found him very hard to work with, and that's being kind! I got requests from his assistant: "David Lee wants the studio decked out like he's at the beach [with] palm trees, real ones." He wanted a red carpet going into the studio with satin stanchions, like the ones at the movie theaters. He wanted a table set up outside the studio with all kinds of breakfast food on it. I said, "Okay, what does he want to eat?"
>
> "No, no. It's not to eat," the guy says. "It gets him in the mindset of breakfast and mornings. He wants Pop-Tarts®, he wants eggs, frozen waffles, McDonald's breakfast; put them all on the table." So the food sat out there, and nobody could touch it. One morning after his show, he was supposed to meet all the sales people in the conference room; they were all waiting in there. He came to my office and said, "I'm not going down to the conference room. I'm leaving."
>
> "Uh, okay . . ."
>
> "I don't go into conference rooms; I'm not meeting anybody." Then he turned and walked out. He did do an event at the Paradise Theater for us where he came out and met people, but then he wouldn't sign anything; not even a guitar for charity.

Mark Hannon later told Joanna Weiss at the *Boston Globe* that the WZLX tryout "was raw Dave. It was no guidance, no direction, no producers, no guests, no music. Everyone who listened to that said, 'Imagine this guy in a setting where he's going to be surrounded by great radio people.'" But the consensus among the WZLX DJs, including myself, was that David Lee Roth was a born loner, surrounded by acres of ego; he would never succeed as a morning show host unless he learned how to operate as part of a team. Nevertheless, CBS Radio announced that on 3 January 2006, Van Halen's former singer would begin broadcasting from New York as the new morning show host for WBCN, along with six other stations. Roth told the *Boston Globe*, "'I'm like a tomahawk missile right in the Cheerios." When the big day arrived, Juanita had set her alarm clock: "I loved DLR; I wanted to hear that historic moment. I listened to the first break and it was . . . crazy! It went on and on and on; what was he talking about? 'This is going somewhere, right?' I knew right then that it wasn't going to go well."

"It's a tough job [being a morning show host]," said Hardy. "It's a craft; you can't just walk in." Five weeks after Roth's debut, the *Boston Herald* confirmed that the program was in trouble: "David Lee Roth's radio days are numbered. Critics pan him. His ratings are horrendous. He's difficult to work with and on air he's dared his CBS bosses to fire him." The story also revealed that WBCN's morning ratings had plummeted from first place to twelfth.

Rumors began swirling that CBS was in secret talks with Opie and Anthony, the tarnished white knights, to replace the failing morning show. It turned out to be true: the company canned David Lee Roth, paying out the remainder of a reported $4 million deal to unload the rock star, and announced a unique arrangement with O & A, who were happily working at premium salary levels for XM Satellite radio. XM would license a three-hour segment of the duo's show to CBS, which would simulcast it on terrestrial radio and edit out any offensive sections. When O & A returned to the airwaves of CBS Radio in April '06, one of the duo's first guests was the chairman of CBS Radio Joel Hollander, who underscored the company's priorities by mandating simply, 'Get me ratings. That's it.'"

"There were decisions made, like David Lee Roth, that were so monumentally catastrophic that any chance the station had of surviving was killed," Steve Strick said. By that time, the veteran BCN jock, newsman, and programmer had already exited, leaving the station for a post at *Radio and Records* magazine in Los Angeles. Wellington promoted overnight jock Dan O'Brien from sister station WBMX to replace Strick as music director.

Then, in June '06, Wellington brought in a new afternoon tag team that Tony Berardini (in his new role as talent head hunter) had located: Fred Toucher and Rich Shertenlieb, whose show emulated O & A's in its coarse style and subject matter.

"After Howard left, the Roth experiment, and the change in the mornings to O & A, consistency was the problem," Hardy observed. "Everybody used to know that Charles was on six to ten in the morning or that Chuck Nowlin was on WZLX in the afternoon. You knew they were on at the same time, at the same station, for years. That consistency breeds the familiarity, and familiarity breeds the partiality, and that gives you your ratings. At 'BCN [in its final years], there was never anything in place for long enough to give out a real sense of what the station was supposed to be."

"I disliked how rigid and bland it was getting," added Mark Hamilton. "The creative brushstrokes attempted were not the first times anybody else had done them; there just wasn't a lot of risk taking."

"[Dave Wellington] took a machete to the music library and cut it in half," moaned Adam 12, "plus the music became sooooo . . . vanilla." By the late summer of '07, Shred had had enough; the 'BCN veteran exited after twenty years on the air and managing thirteen Rock 'n' Roll Rumbles. On his way out the door, Shred gave Wellington a parting shot: "I sent an e-mail on a Sunday and basically told him to 'f' himself. It was an open letter; I cc'd everybody. I torched it, no doubt about that! But then, Wellington was let go, not six months later."

Wellington, however, was not without his successes. Juanita pointed out, "He was the one that brought Toucher and Rich to 'BCN and that show was a huge success." Adam 12 had to agree: "They were responsible for 'BCN lasting another two years. I think corporate would have axed the station if they hadn't shown up." Despite the building ratings of the new afternoon team, the sales reports told a grimmer tale. Radio revenue, in general, had taken a huge hit in the post-9/11 advertising world; even so, 'BCN was the number 3 earner in town, taking in $26.5 million in 2005, the last year of Howard Stern's tenure. But twelve months later, with the shock-jock's departure, that figure had plummeted to $18 million, dropping the station into seventh place. In 2007, despite all attempts by management to correct its course, WBCN's annual take fell nearly another half-million dollars. That number sealed Wellington's fate as he departed amicably in June 2008.

Just ten days later, the keys to WBCN were handed to Mike Thomas,

WBCN drive time is all talk: Opie and Anthony meet Toucher and Rich.
(From left to right) Rich, Opie, Anthony, Toucher (with sidekick Crash on
the right). Photo by Janet King.

program director at WZLX, who now had the duty of helming both stations suddenly thrust upon him. That month, the River Rave returned after a four-year absence, and the Christmas Rave would make a comeback as well. These attempts at a local reconnection continued when Opie and Anthony's syndicated show was dropped at 'BCN in December. Toucher and Rich moved into morning drive, Hardy occupied the afternoons, and Dan O'Brien filled a 7:00 p.m. to midnight slot with music. Amidst the changes and cost reductions, WBCN's 2008 revenue numbers came in, but the news was still not promising and not without consequence. With a nearly $2.5 million dollar drop from its previous year, total earnings at the once-mighty WBCN were now being surpassed by WZLX, which had always earned less.

"A great thing that Wellington did do before he left was inviting Charles Laquidara to come back in March 2008 for 'BCN's fortieth birthday," Adam 12 recounted. "I didn't know Charles at all, and having him on was incredible; I was so flattered. He got to pick his own songs, and he sent the last one out to me; it was 'The Last DJ,' by Tom Petty." Laquidara had most of the songs he wanted to play stored in his computer and plugged the laptop into the control room board. "It was funny how it ended," Adam continued, "because he completely forgot he was playing Petty. He unplugged his

Figure 16.5 Hardy (on the right) backstage at the Boston
Garden with Lars Ulrich of Metallica, 18 January 2009.
Photo by Rich Shertenlieb.

computer and the song just stopped, right in the middle of 'The Last DJ.'
Silence. Was it a foreshadowing of things to come? I guess so."

"The rumors had been around since Howard Stern was getting ready
to leave," Hardy said.

"Even so," added Juanita, "we were still shocked when it actually happened."

"Suddenly, one day, some workers came around and took the WBCN
letters, stenciled on all the windows and doors, off the glass," Hardy said.
"They replaced them with 'CBS Radio.'"

In July 2009, CBS announced that the company would be closing down
WBCN on 13 August. Depending on whom you talked to, WBCN's "American
Revolution" had ended for them in a hundred different possible places
during the previous four decades, but now it was officially being laid to
rest. "It was almost a mercy killing," Nik Carter observed. "There was definitely a sense of, you tuned in and you didn't know what you were going
to get. As much as that might sound inviting, people are creatures of habit
and they go for certain things for certain reasons. Unless you have a really
well-defined reason for being there, what's the point?"

"It was really like your local library closing," Neal Robert said. "If you
asked people when the last time they listened to WBCN was, most would
say, 'Well, I don't listen to it.'

'But are you sad it's closing?'

'What!? They're closing 'BCN?' It had gone from being a necessity to being an afterthought."

In a somewhat complicated arrangement, 'BCN would give up its frequency of 104.1 to WBMX-FM, the CBS station at 98.5, giving "MIX" a much better signal in the Boston area. Launching from the same studios as the defunct WBCN would be a brand-new sports talk station dubbed "The Sports Hub," at WBMX's old frequency of 98.5. Designed as an FM competitor to long-standing Boston sports station WEEI-AM, "The Sports Hub" (official call letters WBZ-FM) built on the benefits of WBCN's long association with the three-time Super Bowl–winning Patriots. Toucher and Rich, already a proven success, would continue their run in morning drive with an added emphasis on sports commentary and humor.

Although relatively new to the market, the inheritor of "The Rock of Boston" (and indeed the new "Sports Hub"), program director Mike Thomas, respected 'BCN's forty-one-year legacy enough to propose a four-day farewell broadcast, inviting the station's long list of personalities to play music from all the different eras, offer up choice observances from the past, and say their goodbyes. In respect to the last Class of WBCN, Adam 12 and Hardy would get to do their own regular shows on the final day. Longtime 'BCN production director John Reilly assembled collages of audio memorabilia, including best bits from standout broadcasts, epic interviews, choice station IDs, and hilarious comedic episodes. The material spanned the distance between Bruce Springsteen's first visit to 'BCN in 1973 to the Dropkick Murphys crashing Nik Carter's show in 2002 and Adam 12's interview with Howard Stern on 'BCN's fortieth birthday. Any historical audio stamped with greatness that could be found was rushed into Reilly's production studio to be organized and readied for a last appearance on 104.1. Invitations went out to WBCN's DJs and alumni to return and share their memories and impressions before the forum to do so vanished. As everyone received the grace of a final farewell, the response was nearly unanimous.

A directive came from the head office, addressed in a final memo to the staff, to curtail any degree of on-air negativity, "bitching," or standing on a soapbox to complain about the reasons such a Boston institution would shutter itself. But the corporate powers needn't have worried; the time to complain about WBCN's demise had passed. Like popular tastes and culture, the station was a moving target that had already entered the history books. The occasion called for respect and recollection; grumbles and grievances

Mike Thomas is handed the keys to WBCN, but its fate is already sealed.
(From left to right) Carter Alan, Dan O'Brien, the Edge from U2, and Thomas,
March 2009. Photo by Leo Gozbekian.

would merely tarnish the memories and distract from the greatness of
forty-one innovative years. Whatever animal the station had become to
any listener, whether they still enjoyed 'BCN or abandoned it years earlier,
was forgotten in the last four days as 'BCN's jocks told their stories and
selected their music the old-school way—by whatever suited the moment.

Sam Kopper, WBCN's first program director, returned to reminisce with
Steve Nelson, early manager of the legendary Boston Tea Party, about the
days leading up to the creation of WBCN in 1968 and the effect the station
had once it dropped into the scene. Generations met generations: with
eyes agape, late-term DJ Mark Hamilton manned the controls for Kopper
and Nelson, drinking in every revelation, and then hosted a one-time only
WBCN "supersession" with Kopper, Charles Laquidara, Norm Winer, Tom
Couch, and Danny Schechter. Adam 12 chatted with Tank, and Hardy welcomed me into the studio to speak about my years of being music director
and 'BCN's key involvement in the careers of artists like Led Zeppelin, Bruce
Springsteen, Stevie Ray Vaughan, and in a personal case, U2. Steve Strick
checked in on the phone from LA and then Chachi took a seat to help field

a suitably ridiculous call from Duane Ingalls Glasscock, who attempted the impossible task of listing in one hour, by name, every person who had ever worked for WBCN.

Oedipus spoke of the cultural edge his 'BCN tenure embraced and shared, from Sex Pistols attitude to the astonishing vocal agility of Jeff Buckley, before hosting a final edition of "Nocturnal Emissions." The last episode of WBCN's local-music haven, "Boston Emissions," united a few generations of hosts including Albert O, Shred, Hamilton, Juanita, and Anngelle, who all enthused about the history and importance of the area's homegrown rock and roll scene. Maxanne, Billy West, Joe Rogers, Ken Shelton, Jerry Goodwin, Tami Heide, and Lisa Traxler offered glowing tributes. Jonathan Kraft called in to express his sadness over the station's passing but praised the Patriots' continuing association with CBS. Flexing his hometown roots as a struggling Boston comic, Jay Leno also took the time to reminisce and thank the station for being an early catalyst to the New England comedy scene.

A tape of Laquidara talking to Al Pacino; Hardy's irreverent and streetwise co-conspirator Robbie Roadsteamer jawing on and on about life with and without "The Rock of Boston"; the rare single "Psycho" by Jack Kittel; a 1979 interview with an unknown John Cougar; Oedipus telling Bradley Jay that his secret was not hiring good people and letting them do their jobs but hiring the *best* people; Mission of Burma's "That's When I Reach for My Revolver"; a discussion concerning Timothy Leary's contribution to culture; "21st Century Schizoid Man" from King Crimson; a seventies-era John Lennon station ID; Bono expounding in a 1997 interview from Foxboro Stadium; and even the station's very last commercial—Gene Simmons hawking Dr. Pepper—all made the four-day farewell a mandatory listening event. Musical moments fit for WBCN's free-form legacy included Hamilton's parting "We'll Meet Again" by Johnny Cash; Juanita's last song, "That's Life" from Frank Sinatra; "Farewell Ride" by Beck; Jimmy Cliff's "The Harder They Come"; "Around the Dial" by the Kinks; "Breathing" from Kate Bush; and Adam 12's parting "I Shall Not Walk Alone" by Ben Harper.

Bradley Jay volunteered for, and received, the momentous job of closing WBCN down, completing a chain that began when Joe Rogers had lugged a box of records up 'BCN's long stairway at its first Newbury Street location, his feet unknowingly initiating a forty-one-year string of DJs arriving for their shifts, plugging in their headphones, creating magic for as long as the schedule gave them, and then making way for the next guy or girl, gay or straight, black or white, stoned ragamuffin or alert teetotaler. It didn't matter; they all had the best gig in the world: a headful of ideas and the vehicle to use them. Now Bradley Jay arrived with another load of "records." Of course, now they were all digitally compiled in a computerized storage platform, instantly accessible as fast as fingers could locate them on the screen. Although this new-world DJ didn't need to worry about a hernia as much as Rogers had, he still lugged as heavy a weight of tunes around in his head. In four days of celebration, the license to freely use those songs and express any thoughts that came to mind had been reestablished. WBCN came around full circle, and the days of celebration became days of closure for the staff and for Boston.

Jay's choice for the final song, "Shine on You Crazy Diamond," originally an accolade to Pink Floyd's fallen acidhead bandmate Syd Barrett, was now transposed to describe the loss of another, also slightly deranged, (radio) icon. The lyrics uncannily fit their new context while the dreamy fourteen minutes of music allowed time for those listening to ponder the station's imminent end. Slowing fading, the song gave way to a brief parting collage of 'BCN voices and fleeting moments before the late Darrell Martinie had the last word. Snipped from the closing of his "Cosmic Muffin" reports, a simple "Over and out!" abruptly marked the end of WBCN's reign, before static and a WBMX station ID previewed the beginning of a new and very different era. "Nothing lasts forever, especially rock and roll," Jay later said of the final moment. "To think that we were going to go on forever is just the antithesis of what rock is. So, goodbye to WBCN; it was just my great honor to be involved."

Wednesday, 12 August 2009: It's the morning after WBCN has gone silent. I push on the main air studio door at 7:00 a.m., expecting to stick my head in on a bustling morning team from the new sports talk station. But I forgot: tomorrow is the day that '98.5, "The Sports Hub," will occupy this studio while WBMX assumes WBCN's former 104.1 frequency. This is a quiet day of preparation before the two stations erase any remaining vestige of "The Rock of Boston." The room is empty, while a recorded message alerts the listeners to the impending arrival of the "hot adult contemporary" sounds of 'BMX. I'm actually relieved; I wasn't really ready to socialize with the folks that were taking over WBCN's turf. I'm here merely to locate two CDs I lost when I did my four-hour chunk of WBCN's farewell broadcast. The control board is idle, the silence deafening. Posters and photos tacked to the walls testify to past glories. On the long coffee table, made from one of WBCN's bumper-sticker-covered studio doors, are piled haphazard stacks of CDs with some empty beer bottles in between. I find my missing discs fairly easily and turn to leave. But I can't. No one is asking me to go; indeed, there's not another soul on this entire basement floor of the CBS building. All of the WBCN people are gone, everyone who came through the station's doors for over forty-one years.

The air is still and heavy in here. It's the weight of thousands of hours of history: the hundreds of bands that visited and the tens of thousands of songs played—from vinyl to digital, "I Feel Free" to "Shine on You Crazy Diamond." It was from a WBCN studio that many heard about the end of the draft and the Vietnam War, the sobering announcement from Matt Schaffer that the space shuttle *Challenger* had exploded on takeoff, or that U.S. troops were crossing the border into Iraq (and years earlier, clandestinely, into Cambodia). This was the place where Mark Parenteau had hosted a young Adam Sandler and Jerry Seinfeld in the same interview years before their fame; where the fictitious April Fools' Parade emanated; where Bruce Springsteen performed; where John Belushi clowned; where the members of U2 played guest DJs; and where everyone from Abbie Hoffman to Pee Wee Herman had popped in. 'BCN watched Peter Wolf earn his wings, Danny Schechter tool his craft, Eddie Gorodetsky and Billy West initiate comedy empires, Maxanne launch Aerosmith and the Cars, Oedipus mother-hen punk rock along, Tony Berardini introduce Metallica, and Charles Laquidara preview the "Morning Zoo" approach that served as the model for just about every wake-up show on American radio and TV. It's the home of psychedelia from Tonto's Expanding Head Band to Radiohead, soul brother Al Green to bluesy jammer Dave Matthews, Desmond Dekker to Matisyahu, Iggy Pop to Scott Weiland, The Who to Pearl Jam, Grace Slick to Alanis Morissette, and Captain Beefheart to Beck. 'BCN's studio heard and played all of these artists and several thousand more in forty-one years. Now, all is calm, like a midnight to six in the morning shift, at about 4:00 a.m., with all the in-studio guests and hangers-on departed and the next DJ on his way. But this time, there's no one coming in. It's over.

Sam Kopper continues the fight. Using a radio channel made possible by the digital technology that allows CBS to broadcast two additional signals on the same frequency, he is the catalyst that helped launch "Freeform-BCN." A potential listener does need to purchase a new HD radio and tune in WZLX-HD-3 to hear the station, but anyone with a computer can pick up the online stream by logging onto wbcn.com and following the link.

August 2009. WBCN's previous home on
Boylston Street becomes a memorial. Photo
by Tom O' Keefe, courtesy of Bostontweet.

"FreeformBCN" debuted as an automated digital station soon after WBCN's demise in 2009, but since then familiar voices from the past including Albert O, Lisa Traxler, and Karen Fox have done live shows. Maybe that's the future, I don't know, but the music is certainly exciting and reflects the spirit of adventure that WBCN always exhibited.

With CDs in hand I leave the WBCN studio for the final time. Maybe Hardy, Dan O'Brien, John Reilly, or Chachi, who have retained jobs on the floor at the new "Sports Hub," will visit after me. In the ensuing weeks and months, the sports talk station will launch with enormous success, building on the back of the Patriots legacy that WBCN pioneered and on CBS's association with the Boston Bruins. Toucher and Rich's morning show, augmented by sports expert John Wallach, grows only larger in ratings success. The station becomes a monster, serving as an example for similar sports-minded radio franchises launching around the country. Meanwhile, if a WBCN listener—who had left the Boston area years earlier and just returned—were to tune his dial to the familiar 104.1, he would receive a severe shock to the system, as the sounds of Madonna and Katy Perry throbbed hotly on the frequency where his favorite station had been. But it has worked out for "Mix 104.1," which now has much better reception in Boston and has found that its audience followed them over from its old frequency without any problem.

Tony Berardini still works for CBS, performing various tasks that we jokingly refer to as "Black Ops," since we never know what he's up to. Oedipus is happily retired, and the rest of the WBCN crew scattered to the wind. At classic rock sister WZLX, I find myself in the unexpected position of being one of the last on-air survivors from 'BCN's glory days, but Bradley Jay is still here, too, at 'ZLX and WBZ. Peter Wolf, Joe Rogers, and "Master Blaster" all live locally. Ken Shelton hangs out in Chestnut Hill, Mark Parenteau in Worcester, and Charles Laquidara swims next to his home along the coast of Maui. Norm Winer still helms CBS's WXRT in Chicago after more than thirty years, while Tami Heide, Tommy Hadges, and John Brodey all work out on the West Coast. I could continue, for a couple hundred more examples, but time or space doesn't allow for that.

The end of WBCN was a critical blow to the exposure of new rock music, which has always been all about expression of freedom and safety from convention. But nearly three years later, the dissolving of WFNX's alternative format after its sale to Clear Channel Communications all but shuts the door on commercial radio's involvement with new rock music in Boston. While WZLX and WROR proudly display big ratings, their formats choose from the best of rock music past, and Radio 92.9 (formerly WBOS), although it spins more recent nineties and new-century sounds, is a jockless jukebox devoted to heavily rotating recent singles. WAAF, long a thorn in WBCN's side by touting a younger and hipper playlist, increasingly abandons the bravado of its past by embracing the heritage it once scoffed at. This is not a good indication of the health of new rock, which has fallen on hard times, not just in Boston, but around the world. Although there are some damn good rock bands out there, they alone cannot generate enough interest among listeners to support a major-market radio station with the ratings it needs to survive.

The daring of rock music and the willingness to promote any means to mutate or evolve it is barely heard these days in Boston. Except for lonely WXRV ("The River"), alternative rock is left to the listener's imagination on commercial radio. We're back to the scene as it existed prior to WBCN's arrival in 1968: a cloud of Top 40 signals surrounding an inner circle of college radio stations willing to expose the latest sounds of rock. It seems astonishing, but Boston radio has closed an enormous forty-five-year loop. Maybe the new counterparts to Joe Rogers, Peter Wolf, or Tommy Hadges are on the air at WMBR, WERS, WMFO, or one of the other noncommercial outlets right now. Perhaps an entrepreneur to rival the hippie ethos of Ray Riepen is devising the means necessary to foster radio's involvement with a new music scene. That would certainly be a great hope for rock music fans as we continue into this new century.

WBCN was more than just one DJ or program director. It was more than the generations of fine news department staff or the sales people who chased the funds that fueled the whole thing. It included the engineers that built or "Mickey Moused" the equipment together on no budgets and the custodial staff that kept the building running. The traffic department that scheduled the commercials and tracked the revenue to the business department that counted the money and paid the bills: they were all part of it. For both the nonglamorous workers to the spotlit celebrities, 'BCN was a living organism, functioning best when it didn't become too self-conscious and just kept moving forward, creating history as it went. Then there were the listeners. You were probably one of them. How far would "The American Revolution" or "The Rock of Boston" have gotten without listeners? For the best audience a radio station could ever have hoped for, this book is really for you. And since nearly every person who ever spoke on the air at WBCN was a fan of the station before they were hired, I guess this is for us as well. The pleasure was all yours and ours. So, goodbye to WBCN, old friend; we now entrust you to our memories. Shine on!

bibliography

Abel, Katie. "Advocacy Journalism: Practice and Peril in Local Broadcast News, 1981–1999." Unpublished essay. 18 September 2010.

Aerosmith. "Support the Strike!" Advertisement. March 1979.

Album Network. Closed Circuit. 26 February 1979.

Alters, Diane. "In Lowell, Cheers and Blisters." *Boston Globe*, 22 November 1985.

Associated Press. "Radio Deregulation in 1996 Spurred Company's Growth." 21 September 2003.

Band Boston. "There's No Place Like Home . . ." Advertisement. February 1979.

Bay Area Radio Museum. "A Brief History of 106.9 FM in San Francisco." Accessed 20 August 2012. http://bayarearadio.org/history/timeline_106.9-fm.shtml.

Bickelhaupt, Susan. "Parenteau's Spirit Enlivens A.I.R. Awards." *Boston Globe*, 6 November 1997.

———. "Schaffer Returns, Jay Departs among WBCN Shifts." *Boston Globe*, 5 June 1997.

———. "Stern Moves to Mornings, Laquidara Moved to 'ZLX." *Boston Globe*, 2 April 1996.

———. "What's 25 and Still Rocking?" *Boston Globe*, 12 March 1993.

Bieber, David L. "Rock-solid Radio." *Boston*, June 1970.

Billboard. "Executive Turntable." 17 January 2004.

———. "Spring '96 Arbitrons." 3 August 1996.

———. "Summer '97 Arbitrons." 25 October 1997.

———. "Winter '97 Arbitrons." 3 May 1997.

Blowen, Michael. "What's So Funny? Supposedly Boston's Comedy Scene Is Not What It Used to Be." *Boston Globe*, 14 December 1987.

Blumenthal, Sidney. "Alternative Radio Signs Off: The Duane Glasscock Flap." *Real Paper*, 12 August 1978.

Bono, The Edge, Adam Clayton, and Larry Mullen Jr., with Neil McCormick. *U2 by U2*. London: Harper Collins, 2006.

Boston Globe. "Ask the Globe: WBCN [sign-on]." 18 December 1980.

———. "Half of WBCN Staff is fired By New Owner." 17 February 1979.

———. "Theodore M. Hastings Jr., 84: FM Radio Broadcasting Pioneer." 1 October 1994.

Boston Herald. "WBCN 104 FM in Association with Jordan Marsh Present the 8th Annual Rock 'n' Roll Rumble." Advertising supplement. June 1986.

Boston Phoenix. "The Cars Support the WBCN Strike." Advertisement. 13 March 1979.

———. Don't Quote Me . . . 6 March 1979.

———. Don't Quote Me . . . 13 March 1979.

———. "15 Incredible Years: WBCN Boston, #1 Rock of Boston." Advertising supplement. March 1983.

———. "104 FM WBCN in Association with Spit/Metro Presents the 5th Annual Rock 'n' Roll Rumble." Advertising supplement. 21 June 1983.

———. "104 FM WBCN in Association with the Boston Phoenix Presents the Second Annual Rock 'n' Roll Rumble." Advertising supplement. 24 June 1980.

———. "104 FM WBCN Rock 'n' Roll Expo '85." Advertising supplement. 16 April 1985.

———. "20th WBCN 104 FM Boston." Advertising supplement. 20 May 1988.

———. WBCN 25th birthday advertisement. 23 April 1993.

———. "WBCN Strike Benefit Shows." Advertisement. 6 March 1979.

Boston Radio Watch (blog). "Boston Radio Saw Big Revenue Declines in 2008." 2 April 2009. http://bostonradiowatch.blogspot.com/.

Boston Radio Watch (blog). "Top 20 Boston Market Stations according to BIA Network's 2007 Revenue Estimates." 2 April 2008. http://bostonradiowatch.blogspot.com/.

Boston Sunday Globe. "Support for WBCN Strike." 25 February 1979.

Boyle, Mike. "O & A Speak Live on New CBS Affiliate." Radio Monitor, 24 April 2006.

———. "WBCN Announces New Line-up." Radio Monitor, 5 June 2006.

Brown, Charles. "WBCN: Has Come a Long Way." Boston Herald American, 17 May 1975.

Cavaluzzo, Laura E. "WBCN Remains Solid as a Rock after 20 Years: Radio Advertisers Find the Boston Station a Sure Bet in a Fickle Market." Adweek, 28 March 1988.

Cobb, Nathan. "Ray Riepen Rides Off into the Sunset into a New Future." Boston Sunday Globe, 12 September 1971.

Craig, Jack. "WBCN Drops Ball in Debut." Boston Globe, 8 August 1995.

Crouse, Timothy. "The Boston Radio War Is Over: Boston Tests New Music and Flunks Out." Rolling Stone, 20 January 1972.

Daily Free Press. "DJ Segal Defends Walk-out." Excerpt of the 20 February 1979 public strike announcement. 21 February 1979.

Davis, Stephen. Walk This Way. New York: Avon, 1997.

Donham, Parker. "Media Freaks Act Out Battles of the Radicals." Boston Globe, June 1970.

———. "Women Liberationists Dump Chickens, Saying WBCN Ad Was for the Birds." Boston Globe, 14 February 1970.

Ebony. "Nelson Mandela Live! A Super Welcome for a Super Hero: Tour of the United States." September 1990.

Evans, Paul. "The Beatles." In The Rolling Stone Album Guide, edited by Anthony DeCurtis and James Henke with Holly George-Warren. New York: Random House, 1992.

Fabrikant, Geraldine. "Two Radio Giants to Merge, Forming Biggest Network." New York Times, 21 June 1996.

Fee, Gayle, and Laura Reposa, with Simone Press. "The Inside Track." Boston Herald, 18 October 2009.

Feinstein, Steve. "WBCN: Stellar Staff Equals Spectacular Success." Radio and Records, 1 March 1985.

Fisher, Marc. "Commercial Breaks, Stretching Far and Wide." Washington Post, 18 July 2004.

Freeman, Kim. "1968–1988: WBCN 104 FM Boston 20th Birthday." Advertising supplement, Billboard, 2 April 1988.

Garboden, Clif. "WBCN 1968–1988: When Radio Was Radical." Boston Phoenix, 3 June 1988.

Gelb, Jeff. "Oedipus on WBCN's Resurgence." Radio and Records, 28 May 1982.

George, Peter. "The Concert Network (WXCN-FM)." Archives, Boston Radio. 14 April 1997. http://lists.bostonradio.org/bri/v01/msg00344.html.

Gold, Allan R. "A Disc Jockey Challenges an Oil Company." New York Times, 12 December 1988.

Goldwasser, Noe. "WBCN Staff on Strike." *Record World*, 3 March 1979.

Goodman, Fred. *The Mansion on the Hill*. New York: Time Books, 1997.

Gould, Jack. "Around Country, FM Turns to Rock." *New York Times*, 20 May 1970.

Griffith, Bill. "Tanguay, Sylvester Lineup Up: WBCN Adds to NFL Team." *Boston Globe*, 7 July 2004.

Heine, Paul. "Up Close: A Decade of Dominance." *Friday Morning Quarterback*, 28 June 1991.

Heine, Paul, and Mike Boyle. "David Lee Roth Booted from Radio Gig." *Billboard*, 20 April 2006.

Heine, Paul, Mike Parrish, and Mike Boyle. "Rock Free or Die!" *MQB—Modern Quarterback*, 14 March 1997.

Hinckley, David. "Opie and Anthony Get a 'New' Deal." *New York Daily News*, 6 June 2001.

Infinity Broadcasting Corporation. *Infinity Broadcasting 1986 Annual Report*. New York: Author, April 1987.

"Infinity Broadcasting Corporation: Company History." In *International Directory of Company Histories*. Vol. 48. Farmington Hills, MI: St. James Press, 2002.

Isaacs, James. "Joe Mississippi Harold Wilson Fats Rogers Is Back." *Real Paper*, 30 August 1972.

———. "Staff of Boston's WBCN Strikes over Firings. *Rolling Stone*, 5 April 1979.

———. "Strike Ends at Boston's WBCN-FM." *Rolling Stone*, 19 April 1997.

Jacobs, Pat. "Easy Listening, MOR, or 'Grown-up Music' (Zzzzz . . .)." *Rewind the Fifties*. Accessed 19 August 2012. http://www.loti.com/sixties_music/Easy_Listening.htm.

Jahnke, Art. "The Business of Being Cool." *Boston*, August 1985.

Johnson, Dean. "Music Event Helps WBCN Rave About Ad Revenues." *Boston Herald*, 26 May 2000.

———. "River Rave Debate Rages On." *Boston Herald*, 31 May 2000.

———. "Rock Shock as Oedipus Bolts." *Boston Herald*, 4 June 2004.

———. "Shock Talk." *Boston Herald*, 17 April 2001.

———. "Stations with Young Audiences Get Boost." *Boston Herald*, 18 July 1990.

———. "WAAF's Flash Idea Sparks Trouble." *Boston Herald*, 27 June 1997.

———. "WAAF-WBCN Battle Heats to Flash Point." *Boston Herald*, 27 November 1997.

———. "WBCN Leaves Its Teens with a Birthday Bash." *Boston Herald*, 15 March 1988.

———. "WBCN Makes Bid to Be the 'New Rock' of Boston." *Boston Herald*, 5 May 1995.

———. "WBCN to Cut Longtime DJ." *Boston Herald*, 5 November 1997.

Katz, Larry. "A Fan's Guide to U2 in Boston." *Boston Herald*, 24 May 2005.

Keller, Jon. "WCOZ Remains Rock's Rock of Gibraltar in Boston Market." *Billboard*, 5 September 1981.

Kempski, Kevin, and Greg Reibman. "Rave Reviews: Concert's Success Bodes Well for Rock on the Esplanade." *Tab*, 6–12 June 1995.

Kessel, Martin. "Chronology of the WBCN Transfer." Unpublished manuscript prepared for the Committee for Community Access. 19 February 1979.

———. "WBCN's Boston Sunday Review: An End to Wallpaper Radio." *Radio Waves*, November 1979.

Kifner, John. "The Mandela Visit: Education Is a Mighty Force." *New York Times*, 24 June 1990.

Knight, Michael. "A 3-Week Strike at FM Station in Boston Ends." *New York Times*, 13 March 1979.

Krupman, Errol S., and Jim Felber. "WBCN-FM Ends Three-week Strike." *Daily Free Press*, 12 March 1979.

——. "WBCN Loses Ads as Strike Continues." *Daily Free Press*, 21 February 1979.

Kuczynski, Alex. "Making a Media Giant: The Personalities." *New York Times*, 8 September 1999.

Kula, Geoffrey. "Hatch Shell Show Good, Clean Fun in the Sun." *Boston Herald*, 2 June 1995.

Laban, Linda. "Rave Fest Inspires Rants Too." *Boston Herald*, 26 May 2003.

Lalas, Greg. "Radio Daze." *Boston*, March 2002.

Landau, Jon. "Rock: Goodbye to Norm and J.J." *Phoenix*, 8 December 1970.

Laquidara, Charles. "A Day in the Life, or, The Big Mattress Strikes Back." An essay as reprinted in WBCN's Tenth Anniversary Supplement. *Record World*, 10 June 1978.

Leary, Timothy. *Turn On, Tune In, Drop Out: Original Motion Picture Soundtrack*. Mercury Records 21131, 1967.

Leonard, Devin. "Sirius Fun: Mel Karmazin Finds New Media Is a Lot Like What He Used to Love." *Fortune*, 14 November 2005.

Lexington Minuteman. "Jazz Past, Present and Future with Eric Jackson." 19 January 2010.

Lichtenstein Creative Media, Filmmakers' Collaborative. "Boston Tea Party (1967–1970)." Schedule. the.american.revolution.fm. Accessed 20 August 2012. http://www.lcmedia .typepad.com/theamericanrevolution/boston-tea-party.html.

Local 262 United Electrical, Radio and Machine Workers of America. "WBCN Strike Victory." Press release. 11 March 1979.

Lozaw, Tristam. "U2 Will Follow." *Boston Rock*, January 1981.

Ludwig, Peter. "WBCN Morning Man Laquidara Attacks Apartheid's Shell." *Billboard*, 18 September 1988.

MacRobert, Alan. "Aw Mama, Can This Really Be the End?" *Real Paper*, 3 March 1979.

——. "'BCN Strike Resolution." *Real Paper*, 17 March 1979.

——. "WBCN Strike Update." *Real Paper*, 10 March 1979.

Marotta, Michael. "Boston Remembers Kurt Cobain." *Boston Herald*, 3 April 2009.

McLaughlin, Jeff. "In the Latest Radio Wars, It's Position That Counts," *Boston Globe*, 20 April 1989.

——. "Shelton goes to WBCN." *Boston Globe*, 22 February 1980.

——. "21 Day WBCN Strike Ended." *Boston Globe*, 11 March 1979.

——. "WBCN Strike Enters Second Week." *Boston Globe*, 24 February 1979.

——. "WBCN, Union Miles Apart." *Boston Globe*, 22 February 1979.

McNamara, Denis. "Radio Station Purchases Could Hit $100 Million." *Billboard*, 10 May 1986.

——. "WBCN Competitors Rock the Album Boat: WZLX, WAAF Challenge Boston's Longtime Leader." *Billboard*, 24 May 1986.

McNamara, Eileen. "Some Sad Days for the Media." *Boston Globe*, 8 April 1998.

Milano, Brett. "8 Boston Bands Play Youth Benefit at Garden." *Boston Globe*, 3 December 1988.

Miller, Martin. "Getting Sirius in His Crusade." *Los Angeles Times*, 16 December 2005.

Morse, Steve. "No Raves from Police at Foxborough Rock Event." *Boston Globe*, 29 May 2000.

——. "Roach Clips: The Life and Times of WBCN's Evening Disc Jockey." *Boston Globe*, 29 March 1979.

——. "Sartori Calls an Exciting New Tune at WBOS." *Boston Globe*, 8 June 1983.

——. "WBCN Rocked by Ratings Slide." *Boston Globe*, 13 January 1989.

MTV News. "Limp Bizkit's Boston 'Guerilla Gig' Attracts Fans, Police." 3 February 2012. http://www.mtv.com/news/.

The New Encyclopaedia Britannica. Vol. 29, *Macropaedia Knowledge in Depth*. "United States of America: Race Relations, Social Changes." Chicago: Encyclopedia Britannica, 1992.

New York Daily News. "3 Busted in St. Pat's Sex Stunt." 16 August 2002.

O'Brien, Dave. "On Strike: Struggle 70s Style at BCN." *Boston Phoenix*, 27 February 1979.

Pierce, Dave. *Riding on the Ether Express: A Memoir of 1960s Los Angeles, the Rise of Freeform Underground Radio, and the Legendary KPPC-FM*. Lafayette: Center for Louisiana Studies, University of Louisiana at Lafayette, 2008.

PR Newswire. "Infinity Enters into an Agreement to Acquire WUSN-FM Chicago, WZLX Boston and WZGC-FM Atlanta." 17 August 1992.

Radio and Records. "Bands Buy Support Ads as WBCN Dispute Continues." 9 March 1979.

——. "WBCN Staff Strikes against New Owner." 23 February 1979.

Ramirez, Anthony. "Radio Giant Is Set for Growth Spurt." *New York Times*, 18 September 1995.

Reelradio. "'Fantasy Park' Demo: KNUS Dallas 1975." Beau Weaver Collection. 21 August 2012. http://www.reelradio.com/bw/index.html.

Reilly, Adam, and Mike Milliard. "Corporate Parent Kills WBCN." *Boston Phoenix*, 17 July 2009.

Romandetta, Julie. "9 of the Hub's Best Strut Their Stuff at Garden Party." *Boston Herald*, 4 December 1988.

Santosuosso, Ernie. "The Beautiful Radio People." *Boston Sunday Globe*, 23 March 1969.

Schatz, Aaron, ed. "30 Years of WBCN: Collector's Edition." *Virtually Alternative*, December 1998.

Schechter, Danny. *The More You Watch, the Less You Know*. New York: Seven Stories Press, 1999.

Shewey, Don. "Boston's Striking WBCN-FM Employees See Support Increase." *Billboard*, 17 March 1979.

Sigler, Jeremy. "Perry Henzel 2001." *Index Magazine*. Accessed 21 August 2012. http://www.indexmagazine.com/interviews/perry_henzell.shtml

Simon, Clea. "At 'BCN, Bad Boy Act Carries into Afternoon." *Boston Globe*, 8 June 2006.

——. "'BCN, 'AAF Ready for Rock 'n' Roll Ratings Rumble." *Boston Globe*, 26 October 2006.

——. "Change Is in the Air at WBCN." *Boston Globe*, 10 June 2004.

——. "New-Rock 'BCN Reverts to Old Ways of Recruiting Males." *Boston Globe*, 14 June 2000.

——. "Oedipus Stepping Down from WBCN Post." *Boston Globe*, 4 June 2004.

——. "WBCN Names Oedipus's Successor." *Boston Globe*, 9 July 2004.

——— . "WBCN Rocks into Brighton Digs." *Boston Globe*, 7 April 2005.

——— . "WBCN's Oedipus Spins More Dials." *Boston Globe*, 4 September 2002.

Simon, James. "Boston Rock Station Listens to Audience to Capture Market." *Hartford Courant*, 6 September 1981.

Smolian, Steven. "WNCN Management History." *WNCN New York 104.3 FM Classical Radio at Its Best 1956–1993*, 30 January 2003. http://www.wncn.org/.

Sonzski, William. "The Cosmic Muffin." *Boston Globe Calendar*, 29 March 1979.

Spurlin, William. "Concert Network and a Terrible Crash." *Bill Spurlin's Blog*. 9 November 2010. http://bspurlin.wordpress.com/wbcn/concert-network-and-a-terrible-crash.

——— . "Maxanne Sartori." *Bill Spurlin's Blog*. 30 October 2010. http://bspurlin.wordpress.com/maxanne-sartori.

Stark, Phyllis, and Eric Boehlert. "FCC Ownership Edict Generates Flurry of Deals." *Billboard*, 14 November 1992.

Stark, Phyllis, Eric Boehlert, and Carrie Borzillo. "Howard in Beantown." *Billboard*, 13 March 1993.

Sullivan, Jim. "Can You Trust a Radio Station over 20?" *Boston Globe Magazine*, 27 March 1988.

——— . "Goodbye Charlie." *Boston Globe*, 6 August 2000.

——— . "On Top in Rock: WBCN Remains the Station Others Try to Beat." *Boston Globe*, 12 January 1986.

——— . "Reinventing the Rock of Boston." *Boston Globe*, 5 May 1995.

——— . "Season's Rocking: WBCN Rave Fills the City." *Boston Globe*, 7 December 1995.

——— . "Static on Rock Radio." *Boston Globe*, 30 May 1997.

——— . "Uncle Betty to Battle Seka for Rumble Honors." *Boston Globe*, 20 May 1991.

——— . "WBCN's Nik Carter Finds Himself in a Radio Ruckus." *Boston Globe*, 28 November 1997.

——— . "Without Opie and Anthony, WBCN Focuses on the Music." *Boston Globe*, 26 September 2002.

Sutherland, Sam. "Studio Track." *Billboard*, 11 March 1972.

Symkus, Ed. "Rock of Ages." *Tab*, 15 March 1988.

United Church of Christ, Office of Communication v. Federal Communications Commission and United States of America, No. 84–1239. U.S. Court of Appeals, District of Columbia Circuit, Brief. 779 F.2d 702, 250 U.S.App.D.C. 312. Argued 26 September 1985. Decided 20 December 1985.

Wastler, Allen. "Loafing on the 'Lesbian Couch.'" Wastler's Wanderings. *CNN Money*. 23 August 2002. http://money.cnn.com/2002/08/23/commentary/wastler/column_wastler/index.htm.

WBCN. "Bill Murray on WBCN with Ken Shelton." Audio recording. 17 June 1981.

——— . "Industry Peers Honor WBCN at 3rd Annual Gavin Awards." Press release. 1 March 1988.

——— . "Local Rock 'n' Roll Comes Alive on Sunday Nights on the Rock of Boston." Press release. 9 February 1989.

——— . "Maurice Lewis Biography." Press release. 20 June 1990.

——— . *On Strike! To Save WBCN-FM*. Strike announcement flyer. February 1979.

———. "10,000 Fans Swarm Boston Garden for WBCN's 'Rock of Boston' Concert." Press release. 1 November 1990.

———. *31st Annual WBCN 104 FM Rock 'n' Roll Rumble*. Official guide. April 2009.

———. "Tony Berardini Selected Vice President/General Manager of KROQ-FM, Los Angeles: Will Also Continue as Vice President/General Manager of WBCN-FM, Boston." Press release. 13 February 1987.

———. "WBCN Acclaimed in *Rolling Stone* Magazine's Reader's Poll." Press release. 9 March 1990.

———. "WBCN and Its Advertisers Ready to Rock at the Third Annual Rock 'n' Roll Expo '86." Press release. 10 April 1986.

———. "WBCN and the New England Patriots Sign Three-Year Broadcast Contract." Press release. 17 November 1994.

———. "WBCN Buys Out Centrum, Gives Away 13,000 Aerosmith Tickets for 18th Birthday Bash." Press release. 15 February 1986.

———. "WBCN Drops the Big One: 329 Pound Pumpkin Symbolically Smashes Multiple-Sclerosis." Press release. 2 November 1989.

———. "WBCN Drops the Big One Again: 415 Pound Pumpkin to Decide Governor's Race." Press release. 30 October 1990.

———. "WBCN Gears Up for Second Annual Rock 'n' Roll Expo." Press release. 12 March 1985.

———. "WBCN Gives Its Listeners Double Trouble: Special Free Stevie Ray Vaughan Concert." Press release. 13 February 1987.

———. "WBCN Gives Its Listeners the Presents . . . 13,000 Aerosmith Tickets." Press release. 28 April 1986.

———. "WBCN Goes Commercial-Free for a Free South Africa." Press release. 18 November 1985.

———. "WBCN Pumpkin Drop Live Broadcast." Audio recording. 31 October 1989.

———. "WBCN Recreates WKRP Turkey Drop." Press release. 9 November 1990.

———. "WBCN Rocks into Spring with 30,000 Listeners at 3rd Annual Rock and Roll Expo." Press release. 30 April 1986.

———. "WBCN Reunites the J. Geils Band and All Former and Current Employees at 25th Birthday Reunion Party." Press release. 8 March 1993.

———. "WBCN to Broadcast Live Mandela Freedomfest." Press release. 6 June 1988.

———. "WBCN Turkey Drop Live Broadcast." Audio recording. 21 November 1989.

———. "WBCN Turkey Drop Photos." Press release. 7 December 1989.

———. "WBCN Van in Death Plunge before Thousands of Listeners." Press release. 3 April 1990.

———. "WBCN Voted Best Station of the Decade." Press release. 4 January 1990.

———. "WBCN's Boston Sunday Review Wins Award for Vietnam Segment." Press release. 9 April 1990.

———. "WBCN's Darrell Martinie, the Cosmic Muffin, Named Astrologist to the Commonwealth of Massachusetts." Press release. 22 October 1993.

———. "WBCN's DJ Charles Laquidara Receives Award for Divestment Commitment." Press release. 28 June 1990.

———. "WBCN's Maurice Lewis Honored by Mayor's Office for Hosting Anti-Drug Video." Press release. 5 April 1990.

———. "WBCN's Maurice Lewis Receives Highest Citizenship Award from Mayor's Office." Press release. 12 April 1990.

———. "WBCN's Rock of Boston Rocks Boston Garden with 9 Bands." Press release. 16 October 1989.

Weiss, Joanna. "Roth to the Rescue." *Boston Globe*, 3 January 2006.

WGDR. "Alternative Media Project." Radio archives, Goddard College's Community Radio Station. 20 August 2012. http://wgdr.net/wgdrsite/archive/AMP/index.html.

"WHCN (W1XPW)." *Hartford Radio History*. Accessed 20 August 2012. http://www .hartfordradiohistory.com/WHCN__W1XPW_.html.

Whitburn, Joel. *Pop Annual 1955–1982*. Menomonee Falls, WI: Record Research, 1983.

———. *Top Pop Albums 1955–1992*. Menomonee Falls, WI: Record Research, 1993.

Wiser, Brian, and Sheila Warren, with Mimi Fox. "Peter Wolf Historical Biography." *Peter Wolf*. Accessed 20 August 2012. http://www.peterwolf.com/.

Wohlman, Garrett A. "Broadcasting in Providence since World War II." Archives, *Boston Radio*. Last updated 17 October 2011. http://www.bostonradio.org/essays/providence.

Wolf, Peter. "Muddy Waters and Me: Adventures in the Blues Trade." In *Martin Scorsese Presents the Blues: A Musical Journey*, edited by Peter Guralnick, Robert Santelli, Holly George-Warren, and Christopher John Farley. New York: Amistad, 2003.

Wolf, Peter, and the J. Geils Band. "An Open Letter from Peter Wolf and the J. Geils Band. Re: The Strike to Save WBCN-FM." Advertisement. 24 February 1979.

Wolmark, Alan. "T. Mitchell Hastings Reflects on WBCN's 10th." Special supplement: WBCN's Tenth Anniversary. *Record World Magazine*, 10 June 1978.

Wood, John B. "WBCN: Somewhere between Artistic Freedom and Anarchy." *Boston Sunday Globe*, 20 May 1973.

index

Mighty Lunch Hour, the, 152, 175, 180, 186. *See also* Shelton, Ken

Miller, Marc, 116–117, 158–159

Mishegas, 146–147, 151–152, 252. *See also* Big Mattress, the

Mississippi's Soup and Salad, 111

Mittman, Bruce, 276, 280, 286–287, 292

Montgomery, Tim, 67–68, 92–94, 127, 131, 249

Monty Python, 94–95, 186

Morris, George Taylor, 105, 271

MTV, 34, 164–165, 170, 189, 213, 233, 238, 256, 294; Limp Bizkit show, 289–290

Mullaney, John, 232–233

Murray, Bill, 186

Murray, Patrick, 245–246

Music: choosing the playlist on WBCN, 32, 41, 46–49; evolution to "Modern Rock," 254–258; music meetings, 46–47, 198–200, 225, 254–255

Natichioni, Julie, 128

Nelson, Steve, 6–7, 27, 312

Nessman, Les, 215–216. *See also* Drop era at WBCN

news reporting on WBCN, 48–51, 54–59, 121–122, 206–212, 243–246; Reuters Newswire Service, 50

Nicks, Stevie, 227

Nirvana, 238–239

Nocturnal Emissions, 237, 293, 302–303, 313. *See also* Oedipus

Nolin, Carla, 175, 192

O, Albert, 171–172, 175, 196, 218, 225, 227, 293, 296

O'Brien, Dan, 307, 309, 317

Oedipus, 147, 153, 162, 165–166, 174–175, 177, 179–180; becomes program director, 159; CBS Vice President of Alternative Programming, 299–300; with the Clash, 150, 167; on hiring Nik Carter, and the subsequent battle with WAAF, 284–286; with J. J. Burnel of the Stranglers, 150–151; "Oedi's Women in Pain," 225, 290; on Opie and Anthony, 295–296; punk scene, 109, 114–115; retiring from WBCN, 301–303; WAAF/ Parenteau feud, 280–283; WBCN strike, 127, 129, 132, 133–134, 136

Opie and Anthony, 279–282, 286–287, 292, 295–296, 307–309

Orson Welles Theater, 95–96, 136

Palmiter, Leslie, 105

Parenteau, Mark, 104–108, 118–119, 128, 130, 185, 231, 280–283; with John Belushi, 149–151; comedy act with Billy West, 177–178; Peter Frampton incident, 228–229; relationship with Aerosmith, 220; remote broadcasts on WBCN, 202, 190, 212–216; as Scotty Wainwright, 60–61, 119. *See also* Boston comedy scene; comedy on WBCN

Parry, Jim, 29, 38–39, 41–43, 46, 61, 120, 145, 147; coining "News Dissector," 55; producing Peter Wolf radio show, 25–27; WBCN strike, 127, 135

Patriots. *See* sports, on WBCN

Patriots Rock Radio Network. *See* sports, on WBCN

Paul's Mall, 96–99

Perry, Al, 26–29, 43, 49, 59, 78, 85, 87, 89, 100; as "Crazy Al," 26, 75; as general manager, 61, 76

piano drop. *See* Drop era at WBCN

Pierce, Dave, 13, 38

Police, the, 167

politics on WBCN, 51–59, 206–212. *See also* Abel, Katy; Burlingham, Bo; Schechter, Danny

Poole, Rob. *See* Hardy

Progressive Communication Corporation, 124–125 *See also* Hemisphere Broadcasting Corporation

pumpkin drops. *See* Drop era at WBCN

punk scene, 114–115, 154